RESEARCHING TECHNOLOGY EDUCATION

INTERNATIONAL TECHNOLOGY EDUCATION STUDIES

Series Editors
Rod Custer, *Illinois State University, Normal, USA*
Marc J. de Vries, *Eindhoven University of Technology, The Netherlands*

Editorial Board
Piet Ankiewicz, *University of Johannesburg, South Africa*
Dov Kipperman, *ORT Israel, Israel*
Steven Lee, *Taiwan National Normal University Taipei, Taiwan*
Gene Martin, *Technical Foundation of America, USA*
Howard Middleton, *Griffith University, Brisbane, Australia*
Chitra Natarajan, *Homi Babha Centre for Science Education, Mumbai, India*
John R. Dakers, *University of Glasgow, UK*

Scope
Technology Education has gone through a lot of changes in the past decades. It has developed from a craft oriented school subject to a learning area in which the meaning of technology as an important part of our contemporary culture is explored, both by the learning of theoretical concepts and through practical activities. This development has been accompanied by educational research. The output of research studies is published mostly as articles in scholarly Technology Education and Science Education journals. There is a need, however, for more than that. The field still lacks an international book series that is entirely dedicated to Technology Education. *The International Technology Education Studies* aim at providing the opportunity to publish more extensive texts than in journal articles, or to publish coherent collections of articles/chapters that focus on a certain theme. In this book series monographs and edited volumes will be published. The books will be peer reviewed in order to assure the quality of the texts.

Researching Technology Education
Methods and Techniques

Edited by

Howard Middleton
Griffith University, Brisbane, Australia

SENSE PUBLISHERS
ROTTERDAM / TAIPEI

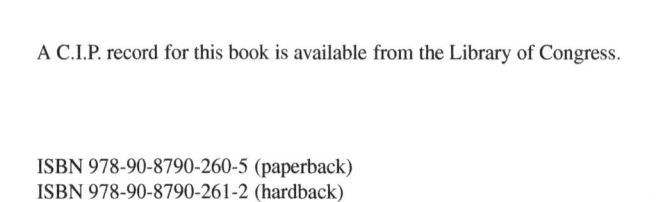

A C.I.P. record for this book is available from the Library of Congress.

ISBN 978-90-8790-260-5 (paperback)
ISBN 978-90-8790-261-2 (hardback)

Published by: Sense Publishers,
P.O. Box 21858, 3001 AW Rotterdam, The Netherlands
http://www.sensepublishers.com

Printed on acid-free paper

All rights reserved © 2008 Sense Publishers

No part of this work may be reproduced, stored in a retrieval system, or transmitted in any form or by any means, electronic, mechanical, photocopying, microfilming, recording or otherwise, without written permission from the Publisher, with the exception of any material supplied specifically for the purpose of being entered and executed on a computer system, for exclusive use by the purchaser of the work.

TABLE OF CONTENTS

Researching Technology Education: New Ways of Understanding
Teaching, Learning and Knowledge   1
Howard Middleton

Classroom Case Studies   6
Robert McCormick

International Collaborative Case Studies: Developing Professional
Thinking for Technology Teachers   28
Frank Banks

The Repertory Grid Technique: Making Tacit Knowledge Explicit:
Assessing Creative Work and Problem Solving Skills   46
Lars Björklund

Researching Expertise Development in Complex Computer Applications   70
Ivan Chester

A Cultural-Historical Perspective on Research in Design and
Technology Education   89
Marilyn Fleer

Design Performance: Digital Tools: Research Processes   99
Richard Kimbell

Comparative Analysis as a Research Method in Technology Education   116
Margarita Pavlova

Observational Techniques for Examining Student Learning Activity in
Technology Education   135
Kay Stables

Capturing Knowledge and Activity   155
John Stevenson

TABLE OF CONTENTS

Using Stimulated Recall Techniques in Technology Education Classes   172
Bradley Walmsley

Examining Design Thinking: Visual and Verbal Protocol Analysis   193
Howard Middleton

Index   213

HOWARD MIDDLETON

RESEARCHING TECHNOLOGY EDUCATION

New Ways of Understanding Teaching, Learning and Knowledge

INTRODUCTION

This chapter provides an overview of the contents of the book, and a rationale for both the idea of such a book and for the chapters within it.

The idea for this book came from a presentation I made at the 2005 PATT Conference in Haarlem, in the Netherlands, which was subsequently published (Middleton, 2006). The theme of the conference, which was in some ways a celebration of twenty years of PATT, was *PATT: Twenty Years in Retrospect*. The intention was to present papers that explored changes in the field of technology education, around the world, over the previous twenty years. Thus, most presenters provided details of the development of the field in their region or country. I took a different approach and provided an examination of the development of approaches to research in technology education over the period. This chapter summarises the argument presented at that time and argued that there had been important developments in how technology education was being researched.

In one sense this book can be seen as the research equivalent of what Shulman termed pedagogical content knowledge (Shulman, 1987). Shulman argued that all teachers needed pedagogical knowledge. That is, knowledge of the general things all teachers need to know, how to structure a lesson, organise class activities, etc. However, Shulman argued that, in addition to these general teaching skills, teachers needed to know the best ways to teach in particular disciplines, the most powerful analogies and examples and ways of creating good learning within that discipline. In this book, I explicitly, and the authors, implicitly, are arguing that for us to understand technology education we need to use research methods that are appropriate for technology education. That is, methods and techniques that account for the particular characteristics of technological knowledge and of the process of learning and teaching in technology education classrooms.

You will also find that aspects of each research method or technique are teased out in more detail that is often the case when reporting on research. This is deliberate and done for two reasons. The first is to assist researchers in making decisions about the use of particular methods or techniques. The second reason is that there is increased competition internationally, for such things as research funding and in getting papers accepted for publication. The assumption throughout the book is that we need to justify research approaches at a more fine-grained level that has been the case in the past. The hope is that the book will help

H.E. Middleton (ed.), Researching Technology Education, 1–5.
© 2008 Sense Publishers. All rights reserved.

in putting research proposals together, both for research students and those seeking funding for research. Finally, it needs to be said that these are not a collection of mutually compatible stories, though of course some are. Some, however, could be interpreted as contradicting the intention or philosophy of others. I prefer to think that there are some tensions across the methods and that this is healthy at any time, but particularly healthy when we are working at developing the best ways of examining this phenomena, about which we still have much to learn, called technology education.

RESEARCHING TECHNOLOGY EDUCATION

In 2005, I argued that:

> The kinds of research methodologies that have been employed over the last twenty years have evolved and are evolving in ways that are making them more suitable for researching the things that need to be researched about technology education. I am not arguing that all research in technology education is compatible with this evolution but that there is evidence that it is happening. My purpose in doing so is based on the belief that using the correct research tools is as important to achieving the research aims for technology education as researching the right topics. Further, some research tools are necessary for the conduct of certain research so that availability of tools can, to some degree, determine what is researched, and what we are able to discover. Lastly, evolution can be ordered or entropic. To ensure that research provides outcomes that allow technology education to evolve in an ordered and positive way it is important to highlight positive developments in research methodologies as well as research findings. (Middleton, 2006, 2)

These new ways of researching technology education included methods that can be considered new, in themselves, such as Richard Kimbell's e-portfolios. Alternatively, some represent revisions to established techniques such as Frank Banks or Robert McCormick's case studies, or Margarita Pavlova's comparative methods. Further, some could be considered methods used elsewhere but not generally used in technology education. These include Lars Bjorkland's repertory grid technique and Brad Walmsley's stimulated recall. However, all are new in the particular ways in which they are used to help us understand knowledge and learning in technology education.

In providing an introduction to each chapter, the intention is not primarily to tell you, in a somewhat condensed form, what you will find in the chapter, although that is done to some extent. The main point is to highlight the features of each chapter that made them distinctive and worth including in the book.

Robert McCormick provides a very detailed description of a case study methodology for researching learning in technology education classrooms. A particular contribution of Robert's chapter is the detailed analysis of the strengths of case study methodology and the detailed description of data analysis. Case study methods, like most qualitative methods, are subject to the criticism that it is difficult to establish validity. Robert's approach to the chapter provides ample material for researchers to use in addressing issues of validity.

In his chapter, Frank Banks extends Robert McCormick's work to detail the possibilities for international collaborative case study methods. One aspect of Frank's paper that should be useful is the ways in which the methodology developed. In an era when greater importance is being placed on research that transcends national jurisdictions, this is a particularly welcome contribution.

Lars Bjorklund revisits Kelly's (1955) repertory grid technique. This is a method rarely seen in technology education research but one that Lars argues is a useful technique to examining constructs about technology education via interviews. Lars argues the technique is useful for tapping into peoples' implicit and tacit knowledge. Given our emerging understanding of the importance of such knowledge to technology education, this should be a useful chapter. Lars introduces us to software that enhances the process for researchers.

Ivan Chester makes the point in his chapter, if only by implication, that when it comes to researching technology education, we are examining a moving target. This seems to be particularly the case with learning and teaching for computer-based aspects of technology education. One issue that has plagued cognitive research into problem-solving processes has been the effect of the research process on the subject and thus on the data being captured. By using more recent data capture software in combination with older verbal protocol capture and analysis, Ivan provides a method for exploring thinking in this rapidly developing area of technology education.

In arguing for the value of a cultural-historical approach to understanding knowledge and learning in technology education, Marilyn Fleer makes the important point that many popular ways of examining concept formation, rely on an examination of formed concepts, rather than the process by which they were formed and the connections between concept formation in school and children's everyday concepts. In doing so she applies Vygotsky's ideas to capture the dynamics of students learning processes.

Richard Kimbell's chapter is truly an action research project given that he wrote the chapter about a project that was still progressing and not due for completion until 2007. The interesting feature of Richard's work on e-portfolios is that they address two long-standing problems in technology education. The first is the problem of conventional portfolios that were intended to describe process (of a design project) which were invariably undertaken after the completion of the project. The second is the issue of capturing, in something near to real time, the students thoughts and ideas as projects develop. This is a method in development but one worth knowing about.

Margarita Pavlova describes a new approach to a well-established research methodology, that of comparative research. However, Margarita brings to the method a rigour and purpose that has often been absent in past comparative approaches, which have often relied on simple descriptive analyses. Margarita does this by subjecting the comparative approach to a strong theoretical framework, which informs both the design of the study and of the analysis of data.

This chapter should provide researchers with ways to undertake international research in technology education that provides important insights into the phenomena being examined.

Observations are a familiar stock in trade for researchers of human activity. However, the important contribution of the chapter by Kay Stables is the detailing of how and why the methods used were developed over time to meet the needs of the situations as they were encountered and developed. Kay also addresses the crucial issue of how to make observations in ways that don't produce completely unmanageable volumes of data for analysis.

The chapter by John Stevenson might be sub-titled a cautionary tale! John draws on research into the activity that occurs in the front office of motels. The chapter is cautionary because John argues from a cultural-historical activity perspective that meaning is not the same as sense and that actions are not the same as knowledge of actions. John goes further in his cautionary tale to make the point that understanding how one person interacts with the artifacts of one context provides limited clues as to how that person, might interact with artifacts in other, apparently similar contexts. John's chapter is a particularly useful one for an area where complex interactions are common, but where there is the tendency to push for simplified, overarching conclusions about the nature of knowledge and activity in technology education.

Brad Walmsley presents us with a technically sophisticated method for collecting, coding and analysing the complex data that comes from the interactions within a technology education classroom. Brad refers to a larger study of which his chapter in this book represents one part. Video Stimulated recall is not new and is a method developed with the availability of inexpensive video recording equipment. The particularly interesting feature of Brad's research is the concurrent capture of both student-to-student interactions and teacher-to-student interactions, and the possibilities these combinations provide for the analysis of the dynamics of the technology education classroom.

My chapter provides a method that attempts to give credence to the particular data that is generated by designing, but which is often ignored by researchers. That is, in the study that the chapter is based on, I used standard verbal protocol analysis of design activity, but added to this an analysis of the visual protocols generated by design activity. The meaningful analysis of visual data is both overdue and one area where there are particularly interesting research possibilities in technology education. It is also an area where research in technology education may result in new insights in other areas where visual data can be generated.

So, the book contains a collection of research methods and techniques that are being used, developed and explored, as the saying goes, as we speak. I hope you find them useful and that they make your research endeavours a little easier and more successful.

REFERENCES

Middleton, H. E. (2006). Changing practice and changing lenses: The evolution of ways of researching technology education, in M J. de Vries, & I. Mottier (Eds.), *International Handbook of Technology Education: The State of the Art*. International Technology Education Studies, Amsterdam: SensePublishers, 53-64.

Shulman, L. (1987). Knowledge and teaching: Foundations of the new reform. *Harvard Educational Review, 19*(2), 4-14.

Howard Middleton
Griffith Institute for Educational Research
Griffith University
Australia

ROBERT MCCORMICK

CLASSROOM CASE STUDIES

INTRODUCTION

This chapter is based on case studies that have been developed over a number of years through examining design and technology (D&T) in secondary school classrooms in England. The focus of these studies was on problem solving, the use of mathematics and science knowledge, and the knowledge demands of electronics. Such case studies are important to capture the pedagogy of the technology classroom and to understand how concepts and processes are made explicit, through the physical and intellectual activities and interactions among students, and between teachers and students.

The theoretical issues of knowledge that these case studies explored concerned firstly the *nature of knowledge* in the technology classroom. Following McCormick (1997 & 1999), knowledge can be seen in terms of general categories of procedural, conceptual, and qualitative knowledge. However, it will be evident that these categories need to be viewed from particular views of learning (McCormick, 2006). In the context of technology it is also important to see the nature of conceptual knowledge in particular, from the point of view of how students experience knowledge through the related school subjects of science and mathematics and how these might differ from that in technology. The second issue concerns the *use of knowledge*, specifically the use of science and mathematics in technology activity. The third issue is to look at some of the slightly different issues of knowledge that might relate to *social or moral issues*, but which occur in the context of apparently technical activities such as programming a traffic light sequence for a pedestrian crossing. The fourth issue is the *teacher's role* in, and the strategies for, dealing with these knowledge issues in the classroom; for example, when teachers come to 'revise' science knowledge needed at the beginning of a technology project, or the extent to which they rely on providing knowledge on a 'need to know' basis.

Some of the studies that this work is based on concern problem solving and the design process, i.e. procedural knowledge. However, in such studies (e.g. McCormick & Davidson, 1996) it became clear that procedural and conceptual knowledge were intimately connected. These studies, and subsequent ones specifically on knowledge, took a situated view of learning with its consequent view of knowledge. This gave particular imperatives to the case studies, and indeed was a fundamental reason for adopting a case study approach. This rationale will be examined shortly, suffice to say that knowledge is seen as intimately tied to the context within which it is encountered (learned), and cannot be seen as de-contextualised and abstract as is usually the stance of cognitive constructivists.

McCormick (2006) shows the general view of this stance and some of the kinds of studies that were undertaken within this framework.

In general these studies were conducted by outsiders (i.e. not by teachers themselves) and this has specific implications for how we conducted the studies and for methodological issues such as the role of the researcher and ethics. (These will be dealt with under *Case Study Methodology*.)

The structure of the chapter follows the major questions that researchers must answer in their decisions and justification for employing case studies, namely why use this approach, what kind of case study, what stance is taken to the major methodological issues such as generalisation, bias, ethics and how data should be collected and analysed?

WHY CARRY OUT CASE STUDIES?

The principal reason for employing a case study methodology is to see a phenomenon under investigation in context. Thus we can set up specific activities to test student understanding of a concept but, unless the activity is encountered as it is in everyday classroom activity, there is no guarantee that it would inform us sufficiently about that everyday activity. This does not preclude such 'tests' as one of the methods of data collection, as I will indicate later. Yin (2003, p. 13) gives a second reason for employing a case study, namely when there are unclear boundaries between phenomena and context.

Both of these conditions are important when research takes the theoretical perspective of a situated view of learning and hence knowledge. McCormick argued that:

> Participation, in the situated approach, means more than that learning is not simply 'in the head', nor just that it takes place in a social *and* physical context, but that it is related to action ... the relationship between thinking and action is reciprocal. McCormick, (2006, p. 33)

McCormick goes on to argue that seeing knowledge as situated implies that it is interwoven with context. A simple example is the concept of resistance, which in a science classroom will be associated with coils of wire, with longer coils constituting greater resistance. In contrast in the technology classroom, resistance will take the form of small ceramic resistors, all of the same size, but differently colour-coded. To understand the distinctions requires the concept of a material's 'conductivity', not usually dealt with in the science lesson at the stage where students will start to use ceramic resistors. In this situation it is unhelpful to give de-contextualised accounts of knowledge and its use. The way teachers are trained, the way the curricula are formulated, and what students are required (even allowed) to learn at various ages, all constitute 'context'. This exemplifies the situation Yin had in mind in requiring both 'real-life' phenomena and an unclear boundary of context and the phenomena.

Seeing knowledge in context is a general rationale for using a case study, but there are also specific purposes in conducting a case study. Up to now I have implied that the research in some way leads to a theoretical understanding. Yin (2003, p. 15) gives five reasons in the evaluation research context, where researchers seek to understand, for example, how an innovation might develop (e.g. take root in a school), so that they can evaluate whether it has been successful and help others implement it in their classrooms. The first purpose is to *describe* an event or an intervention. In our evaluation of a new programme of electronics, we described how teachers used their training, in the many different forms of implementation of the programme in schools, through a number of case studies of their practices in the classrooms (Murphy *et al*, 2004). This also enabled Yin's (2003) second purpose for case studies, namely the function of *illustration*.

From this it was possible to determine messages that emerged about practice and, in the context of this particular electronics programme, the needs that teachers had to implement electronics in their schools. These 'messages' moved us beyond description, to *explore* the outcomes of a real-life intervention too complex to be handled by other research methods (Yin's third reason). We found that for the implementation of an electronics programme facilities, equipment and resources were important, along with local support for a teacher, the curriculum options and so on. The way these factors (and others) played out in a particular case study varied, but as researchers we could show how, for example, effective practice could be encouraged or limited by examining their relationships.

Although this kind of exploratory evaluation study may not lead to theory building, in other studies it did. For example, how students undertake problem solving and of how qualitative knowledge is used by teachers and students in technology classrooms (McCormick, 1999). These case studies explained causal links in real life activities of the classroom. For example, the teacher behaviours that encouraged various strategies of problem solving that students employed. This is not the 'grand' theory that Hammersley (1990, pp. 105-109) discusses when he examines some classic school case studies and 'differentiation-polarization theory', which explains the impact of streaming and the like in schools. However, case studies of all kinds can build on existing ideas and emerging theories.

Often those who undertake case studies for the first time treat them as descriptive, and assume they must keep an open mind, unsullied by theory. However, open as researchers we aim to be, we always bring some kind of 'theory' to bear in trying to understand what we see. Also, it is hardly sensible to start a study assuming nothing is known, and there is no point in undertaking a literature review of theory to put it aside when conducting a case study, assuming that theory will emerge!

All research needs to be based on some kind of research question, and a case study approach must be driven by the nature of that question or questions. Yin (2003, pp. 5-9) gives a useful way of thinking about these questions. If the question is a 'what' question, such as 'What problem solving strategies do students employ?' or 'What science knowledge does a technology teacher assume for electronics projects for 13 year-olds?', then an exploratory case study is appropriate. Descriptions of the science concepts employed will lead to tentative ideas that might be further examined in subsequent case studies. If the question is a 'how' or a 'why' question then the case study needs to be explanatory. For example, 'Why do students use particular problem solving strategies?' or 'Why do students solder a resistor *along* the 'track' of a printed circuit board (PCB), rather than across an appropriate gap in the track?'

Many research projects involve intervening in the classroom. The research question might be 'How can teachers use the science knowledge of electrics that students bring from that subject to technology electronic activities?' It is possible to try out different strategies, in effect to 'experiment'. The amount of control over the 'experiment' will vary but, unless it is a specific teaching approach (e.g. as in the Cognitive Acceleration Programme; Shayer, 1999), it is likely that its implementation will vary according to the teacher, the type of technology project, and a number of other factors. The result is a 'field experiment' or a 'quasi-experimental' study (Yin, 2003, p. 8). A case study approach can be the same, but to account for the nature of the interventions and the results of them are demanding. A teacher researching in her own classroom, could use an action research approach, with cycles of action and research each feeding the other (Kemmis, 1993). External researchers, could use a design experiment approach, where theory-led hypotheses govern interventions and research (Cobb *et al.*, 2003).

WHAT IS A CASE STUDY?

What is its nature?

I have already examined this question in relation to why and in what circumstances case studies are to be employed; that they deal with real-life situations with unclear boundaries of phenomena and context, gives some feel for their nature. Walker defines a case study in terms of it being an:

> '...examination of an instance in action. The study of particular incidents and events, and the selective collection of information on biography, personality, intentions and values allows the case study worker to capture and portray those elements of a situation that give it meaning.' Walker (1986, p. 189)

This definition envisages multiple sources of evidence and, in classroom research, this includes the behaviour of teachers and students. Typically, in an investigation of the use of knowledge in a technology classroom, a case study would include observations of all the lessons in a project (perhaps lasting 6-12 weeks of 3 hours of lessons per week), a collection of the products (e.g. design folios and 3-D models), interviews with the teacher at the beginning and end of the project, student interviews and even small tasks (probes) on their understanding of particular ideas. Student questionnaires can be use to ascertain more general views across a class or a year group. All these give a rich picture of the activity, but work done at home is missed. For example, the main creation of design ideas for students is often done for homework, and all that can be 'observed' is the results of this work (McCormick *et al.*, 1996). This varied data collection allows for the triangulation of data, such that no one source is relied upon to provide all the evidence for the phenomena.

Some of these data (e.g. questionnaires) are quantitative. Case studies don't have to include only qualitative data, often seen as a way of avoiding statistical generalisation; of which more later. Another common misunderstanding is that a case study is an ethnography, and involving participant observation. Ethnographies are close up, detailed observation of the natural world and attempt to avoid prior commitment to any theoretical model; they may not result in case studies (Yin, 2003, p. 14). Case studies similarly don't have to be bound by these conditions for ethnographies. There are similarities when classroom case studies are undertaken by a participant observer; this is but one type of case study. Nor is the use of qualitative data a defining characteristic, as I have already noted.

What is a case?

It is the focus on a case that gives a case study its distinctive character. Again, however, we have a source of confusion, as inexperienced researchers often confuse the *setting* for the case study with the *case* itself:

> A setting is a named context in which phenomena occur that might be studied from any number of angles, a case is a phenomena seen from one particular angle. (Hammersley & Atkinson, 1995, p. 41)

Thus a classroom case study will have the site as the classroom, but the case could be the teacher, a particular student or a group of students. The case of course could be a physical classroom if the focus was on how a particular room aided or not particular kinds of technology activity. In that case the classroom could be observed with different groups of students in it, say, throughout a day or week. More commonly, I have studied a project as the case, by considering the way both teacher and students carry it out. Within this a group or pair of students could be the focus, to understand their experience of the project. It is still possible to follow other students and the overall progress of the class, as part of a project case study. Clarifying the case defines the way data are collected (e.g. what is observed in the classroom) and the unit of analysis for the data.

The classroom as the site does not bound the research, as the study can consider the national curriculum, other sources of views of technology outside the classroom, or indeed the school, which might influence or determine classroom activity. Similarly school or departmental views on, for example, links among departments, can affect how related concepts in the subjects are experienced. In a study of the use of mathematics in orthographic projection in technology (Evens & McCormick, 1997) the teacher asked students to come to the board to add to a developing drawing, and was surprised that a student was able to deal with a hidden element in a drawing (a hole in block). He asked the student 'How do you know how to do that?' to which the student replied 'We learned it in maths sir.' In that school, there was little connection between, the teaching of orthographic projection in D&T and that of 'plans and views' in mathematics.

CASE STUDY METHODOLOGY

In this section I examine methodological issues: quality, through different forms of validity and reliability; the need to minimize bias; sampling; the contentious issue of generalisability; the role of the researcher; ethics.

Quality of research design

Yin (2003, pp. 34-39) deals with quality by applying four tests, those of external validity, construct validity, reliability and internal validity. These cover two of the most important of the methodological issues outlined above, that is, bias and generalisation. He helpfully relates each of his tests to the various elements of the research process (Yin, 2003, p. 34). These are roughly ordered according to the research process elements: research design, conceptualisation of issues, data collection, data analysis and reporting.

External validity

This test requires the researcher to be clear about how the study's findings generalise. A case study is an instance and so in what sense does it have any lessons for other cases? Usually we think about generalisation from the statistical point of view. If we take a sample of students and test their understanding of a technological concept, and have chosen a sample to represent 'all students', then we can have some confidence that, whatever we conclude about this sample, will apply to the population (all students). Of course it is not as simple as that; there are always limitations on the sample and how it will represent all students. The work of Kimbell and his colleagues in the large testing programme of D&T in England, was able to have some confidence in its generalisability to 15-16 year-olds in England because it used large samples (APU, 1991). However, even this robust study had little to say about this age group in Scotland, let alone in other parts of the world; the design processes outcomes for students is dependent on their particular curricular experience. Case studies, in contrast, do not depend on statistical generalisation, rather on *analytic generalisation*, in other words they generalise to a broader theory (Yin, 2003, pp. 10-11). In technology education this is difficult as we have few theories; for example, 'a teacher-created problem solving culture' noted earlier. Walker (1986, p. 191) argues that it is the reader who tests whether it applies in her situation, and is not an author's problem. This argues for a kind of resonance in the findings, either by appealing to teachers' experience or to a theoretical position.

This contrasts with the case study carried out by evaluators, who might take a more 'grounded' approach, not directed by theory. They may not aim to inform future research work, rather the progress of a curriculum innovation. Nevertheless they will implicitly base their work on implicit theories of change or of the processes they are observing in the classroom. In evaluating an electronics in schools programme, our views on gender effects made sense of the way girls were treated in the classroom by a teacher, and how they reacted to particular topics. When discussing a pedestrian crossing, girls would more commonly talk of the needs of old ladies, or those with young children, to have time to get across (this affects the wait time built into the programmed sequence). We were appealing to theories about gender (Murphy, 2006, pp. 227-228).

If there is an appeal to a body of theory, then it should be possible to replicate case studies. Another researcher studying a technology classroom should, for example, be able to see issues of students being unable to use their science learning. Banks (Chapter 2) shows how he and his colleagues moved from a single case study on teacher knowledge and sought to replicate this in different countries, giving this work external validity.

Construct validity

This is an issue in the data collection phase and requires any concepts of relevance to the area of research are operationalised sufficiently to allow data to be collected. Thus if 'uses of mathematics knowledge' was important then this needs to be defined clearly enough to identify its occurrence in the classroom. In the study of students learning orthographic projection, a teacher described the 45° angle used to project a side elevation to a plan view (Evens & McCormick, 1997), but did not use ideas of 'transformation' or 'reflection', understood in mathematics. He dealt with it procedurally, and even spoke of it as a 'magical' process. The result was that when children produced their plans and elevations, some didn't draw the line until afterwards, having taken measurements and manually transferred them, some drew it at another angle and produced incorrect plans, and others did it correctly. Analysing tasks for the potential use of mathematical concepts, and their technological equivalents, enables the classroom observational evidence or interviews of children and teachers to be collected. This range of sources allows another form of internal validity, that of multiple sources of evidence.

Construct validity is at threat when an observational event is being recorded and an inference made; for example, when a student draws the 45° angle and correctly projects the dimensions does she understand reflection? She may be blindly following the procedure the teacher laid down, with no such understanding. If there are no other data to support this inference, then alternatives need to considered or indeed no inference made.

At the global level of analysis, it is possible to feedback to teachers, and students, the emerging analysis, either in the form of a draft case study, or interim findings. They can then give a view on how it reflects their understandings. A study by Logan (2005), investigating the knowledge of an art and design course in a college, started off by analyzing this knowledge in terms of theoretical ideas (e.g. conceptual and procedural knowledge), but when this was presented to the lecturers who conducted the course they were lukewarm about it, causing the researcher to rethink the analysis. Subsequently she re-analysed the observational and interview data around metaphors used by the lecturers, and this resonated very well with the lecturers and gave her more confidence in this analysis.

Reliability

Quantitative analyses use this measure. In structured classroom observation it is important to know the consistency of a person categorizing behaviour; for example, when she repeats the same categorization at different times (intra-judge reliability) and whether two different people observing the same behaviour categorise it in the same way (inter-judge reliability). A case study should be able to demonstrate that the operations of the study can be replicated with the same results (Yin, 2003, p. 37). This is not the replication for external validity, which requires another case study, rather the aim is to minimize error and bias. This requires clear documentation of the steps taken in collecting and treating data. Our studies of problem solving also involved more than one person in each step to make it explicit and open to scrutiny. This was done at different levels. First, the notes for a lesson were written up by the researcher who had been in the classroom, to give an outline of the main elements of the lesson including teacher and student activity. This used a standard format, clearly distinguishing behaviour from comment or inferences about that behaviour. The notes could contain information about specific video-recording decisions taken (e.g. to follow one rather than another, when students were moving around the classroom or workshop,). These notes were used while viewing the video and from this a transcription of the lesson created, giving the actions and associated talk. Figure 1 (p 14) is an extract from such transcription for two students (K & T) who are working on a money collection box where a coin dropping in caused a woodpecker to peck (via a lever mechanism); RW is the teacher. This transcription includes time, what is said, descriptions of activity (in parenthesis when it is associated with talk), guesses at inaudible words e.g. [*pivot*], and comments (again in italics and parenthesis). The documentation and use of a second person at each level of the process helped to minimise the bias of individuals. Incidents that were missed or misheard at the video transcript stage were added.

Other sources of bias relate to the observer. The presence of non-participant observers can lead to bias by the teacher and the students 'performing'. In our experience, even with video recording, this performing effect is shortlived and both teachers and students soon forget the camera. At times students will cover their microphone (if they are fitted with one), in an attempt to prevent the recording of a part of their conversation. One way to minimize such bias is to be unspecific in what they are told is being observed; e.g. say 'I'm observing the way knowledge is used in the classroom'. This is accurate (important for ethics), but not so specific to cue particular behaviours. Another source of observer bias occurs when the teacher asks for feedback on her performance immediately after the lesson. Unless this feedback is part of the research approach (e.g. as part of some action research or collaborative enquiry), it should not be given or ought to be neutral. This needs to be initially clear to teachers, so they know what they are to expect. It does not preclude feeding back what is seen at the end of the data collection phase or indeed as part of the interviewing. Unless particular analyses are being tested (as in the example of Logan's 2005 work), focus the feedback on what was seen, rather than making evaluative judgments about a teacher's performance.

11.17 K puts this strip at the end at right angles to the previous one and pokes a hole through both with a pen.

T: "Use this." She hands K a paper fastener.

K takes it, says something to T who goes to fetch a second paper fastener. K puts the fastener through the two pieces of card.

K puts a fastener through the bird and fastens it to the backing card.

T pulls the two levers backwards and forwards. She adjusts the clip to hold the two together tightly at right angles. *[K has made an L shape for the lever and pendulum. She must be able to see that on RW's model the shape is an off-centre 'T'. Why has she not copied this? Is it because she doesn't understand the function of the weighted short arm?]*

K: "Right now I ...... this then." she turns over the card and takes the levers from T.

T: "You've done it the wrong way. *[all should be on one fastener]*"

K, ignoring T, removes the paper fastener and joins the levers to the back of the card using the fastener that is holding the bird, and hence puts the bird on the pivot.
11.18: T picks up RW's model. She points to the front left hand side of the model: "So, on this one the bird is going to be here."

K: looks but doesn't immediately comment. She turns their model over: "That [*pivot?*] needs to be stronger here. It's got some weight on it. That [striker plate] needs to be cut." (The striker plate protrudes past edge of front plate.)

Figure 1: Extract from video transcript

The situation for participant observer bias is quite different. Participation restricts the way classroom behaviour can be sampled: attention at times being on working with students or generally teaching; being unable to be in the right place and to move to where the action is; just a lack of time to research (e.g. note taking) as opposed to participating. These apply to a lesser extent in video-recording, with bias more likely at the point of analysis, the observation data will be seen through the eyes of a participant. Yin (2003, p. 93-96) notes that bias in participant observation comes through not being in an outsider role, being an advocate, or 'going native'. This source of bias is also the source of insight, and participants are less intrusive than outsiders.

Internal validity

There are threats to validity each time an inference is made at the point of data collection, or later in analysis. In the Figure 1 transcript, the low-level inferences are indicted by italics and in parenthesis. As noted earlier, a second researcher can check the transcript against the video-recording and a second version produced. This transcription can then be put into a framework that indicates the activities and talk of the lesson (Table 1, considered later), and from this a 'lesson analysis' produced. This analysis draws out issues from the lesson, and is again subject to the second person's view. Figure 2 gives an extract from a lesson analysis, where the focus is on the kinds of problems a pair of students encountered in their work (corresponding to the transcript in Figure 1). This lesson analysis is then used across the whole case study, and might involve a re-working of the particular lesson analyses if others in the team disagreed with the way this was being analysed. Each of these steps could give rise to validity threats, and checking them minimises them.

This [extract] is a series of inter-related problems:
- location of bird relative to tree;
- movement of bird;
- rigidity of mechanism;
- effect of coin size on distance travelled;
- guidance of coin to striker plate and catching coin;
- balance mechanism (the lever system);
- effect of pendulum in providing oscillation.

The students each focus on different problems at different times and even when, say, T is talking about one problem K may be talking about another.

Time	Problem posing	Solution giving	Comment
11:16	K asks about location of pivot and then about the pendulum.	(It is not obvious that there are any solutions mentioned.)	Pivot location arises out of T's question on location of bird. Does this mean bird is on the pivot?
11:17	T says K done lever system "wrong way".		Not clear in what way it is wrong; it appears that she is objecting to K using a single pivot, i.e. that bird and lever system are directly connected.
11:18	K diagnoses a problem of construction of the lever system (to take weight).		
11:18	T is more concerned with location of pivot and hence the bird (although this connection is not made explicitly).	T's 'solution' is to refer to RW's model.	The collaboration is at cross purposes ... because T is focusing upon the front view and sees the pivot in relation to that [11:17 comment] and K has seen connection of lever system and bird.

Figure 2: An extract from a lesson analysis (corresponding to Figure 1)

Internal validity is important in explanatory or causal studies. The implied inferences and explanations need to be examined and possible rival explanations explored. For example, two explanations for an analysis that indicates that students are not able to use their science knowledge because:
- a teacher allows insufficient time or opportunity for students to explore their science concepts at the beginning of the lesson;
- or that the teacher is not interrogating students' understanding at appropriate points in their work.

The two explanations are both plausible and incidents need to be examined to see which is better supported or if both are causes.

Becker (1958) suggested that to improve the validity of such explanations it is appropriate to look for negative or counter cases. At the very least it is necessary to be open to contrary findings. This can be encouraged by asking colleagues to look at the findings.

Role of researcher

I have already made points about how different roles give different sources of bias at different times in the research process. An outside researcher assumes an unobtrusive role, though observing a group of students, is likely to cast the researcher as a teacher and they may well ask questions and try to clarify the task or ask for help. Young children treat a researcher as a helpful adult whatever the researcher might tell them. When the researcher asks students questions they will treat him or her as a teacher, or at least think that the responses might be fed back to the teacher. Although participant observation can overcome some of these problems, and is less disruptive of normal classroom activity, there are limited roles for such participation; usually researchers can only be a teacher, not a student (Stenhouse, 1982)!

Ethics

The ethics of classroom case studies does not present many special issues, though anonymity is hard to ensure to those who are in the school, and know that research is being carried out in a particular classroom (Stenhouse, 1982). Other issues include obtaining clearance from students' parents, and making it clear for video that, unless permission is specifically sought, it is not possible to show this video to others. A researcher should not allow dangerous behaviour, racialism or bullying, though whether the researcher intervenes or informs the teacher will depend on the circumstances. However, feeding back what students say and do to their teacher should not usually be done, if the students' confidentiality is to be respected. It is possible to feedback general views of students, or, as noted above, give the teacher general feedback at the end of the research.

Sampling

Assuming the case is a particular teacher or class group, working on a particular project, the choice is to sample lessons (and particular times within lessons), students, events and the choice between a focus on students or teacher. Even in case studies it is sensible to seek representative samples of lessons, students and events etc. We chose projects as the case so that we could see how a complete design task gave opportunities or made demands on students for problem solving or use of knowledge, and we hence observed all lessons of a project. This is very demanding, involving observing three hours per week for up to 12 weeks, and sometimes this amounted to three visits a week. If the research focus is on how teachers prepare students to use technology knowledge, say of 'systems' in an electronics project, then focus on particular lessons where there were some specific 'system' teaching, or where the circuits were being constructed. As students work at different speeds, predicting when they get to particular points later in a project can be difficult; this problem can be avoided by talking to the teacher during the project. This sampling of lessons is selective and driven by 'theory' (or at least to observe particular events of interest). If the research focuses on a teacher's approach to the use of knowledge throughout technology teaching, including non-project based work, then sample across the teacher's activities in a way that is representative of all her activities.

Having chosen a particular lesson then there is the choice of when in the lesson to observe. The simplistic way is to observe it all, but alternatively beginnings, middles and ends of lessons. More difficult is selecting what to pay attention to in a lesson. If the focus is on the questions students ask teachers or teachers ask students, then the teacher and her interactions can be observed. It is not very useful just to say that 'significant events' will be observed, as 'significance' has to be defined. When we researched what problems students encountered and how they solved them, there was no inventory of such problems, nor could we predict when they occurred. Figure 2 shows how we selected them from the lesson, relying on video recording that was analysed after the event.

Use of video does not, however, remove sampling problems. There is still a choice, determined by the research questions, of which students to observe or whether it is the teacher that is the focus. If the students are to be the focus (e.g. to see how they use knowledge), then it is possible to sample students differently:
- in a systematic way, observe each student (or group of students) for a few minutes, an approach best suited to systematic observation;
- selectively focus on particular students (least or most talkative/active; high or low achievers etc.);
- selectively focus on particular activities and follow those students doing these activities (e.g. if they are using circuit testing equipment then observe all students who use it to see how they diagnose and solve problems).

In our work, we tended to focus on one group if it had up to four students or on two groups if they were working in pairs. This allows for absences, inactivity and some variety of behaviour. When we followed project work, then we chose students who the teacher thought were reasonably good, though groups inevitably contained a range. If interviewing is important, choose children who are articulate, unless the quiet child or the low achievers are the focus; interviews that consist of monosyllabic answers can be unenlightening!

THE DATA COLLECTION METHODS

Here I will examine what is available to the technology classroom researcher and consider their advantages and disadvantages. The methods to be considered are: audio and video recording of lessons, including the practical issues of studying and recording students and/or teachers and the mechanics of observation; interviewing; exploring learning outcomes. Of course this leaves out important issues such as collecting documents (e.g. departmental or school schemes of work), lesson materials (e.g. task sheets or that intended to support learning of ideas or techniques or indeed textbooks), and student work. Apart from keeping good records of these and making sure that copies are obtained immediately (rather than requested from the teacher at some later date), there are no special issues for the technology education researcher.

Audio and video recording in lessons

Before considering these specific recording techniques it is worth mentioning the alternatives. One is 'live' systematic observation that gives quantitative data, of the kind produced by Kimbell, Stables and Green (1994). They used a number of categories of activity, including: the intention of student activity (e.g. developing ideas), how it was manifest (e.g. by drawing), the amount of learner engagement, teacher intervention, and what issues the student was dealing with (task, communication or making). They recorded such categories in five-minute periods throughout a lesson, along with a narrative of the activity (i.e. a qualitative account), to make sense of the quantitative data (see Stables & Kimbell, 2006). They took this approach, because they had a clear idea of what they wanted to observe built on a model of the design process.

It is also possible to rely on narrative accounts, or at least accounts that try to record faithfully as much of the activity and talk around it as is physically possible; this results in researcher field notes. In a technology classroom this is very demanding because there is talk, action and artefacts that all inter-relate and recording them simultaneously requires experience and skill. It is probably only possible if there is a focus on one student (or a pair), or a particular kind of event that is easily identifiable.

For these reasons, I prefer video and audio recording of classroom activity. This allows events to be more or less captured, time to reflect on the record, and allows the compiling of specific data (in ways I have already intimated). Video cameras are very good quality, extremely light and small, reducing their impact in the classroom. We started using video in the early 1990s with a large shoulder-carried camera, requiring a cameraperson, and a researcher, which was quite obtrusive. Ten years later, we used a very small camera, with high-quality of recording (digital rather than analogue). The issues for the researcher are, however, still the same, one set relating to the visual image and another to the audio recording.

For the visual image the first question is: where will the camera be pointed? I have already discussed the sampling issue and, if the focus is the teacher, that is straightforward. There remains an issue of where the camera is located and how the teacher is tracked. In a workshop, or an area where both the teacher and the students are moving about, the decision depends on needing to see the details of what the students or teacher are doing. If details are needed, the camera probably has to follow the teacher, which can be obtrusive. If not, then it is possible to put the camera on a tripod in the corner and track the teacher as she moves around. Leaving the camera on a wide angle, unattended, is unlikely to be satisfactory as there will be times when the teacher goes out of shot or is obscured by a student or where it is necessary to zoom in on an activity. This is taxing for a single researcher, as camera work then dominates time in the classroom and taking notes of important events or reactions etc., has to be reduced. My experience doing this in a computer suite (with the camera on a tripod in one corner), where programmable chips and buggies were being constructed, indicated difficulty in recording the program on the computer screen the teacher and students were discussing; subsequently trying to reconstruct what they were talking about was difficult.

Audio recording in this situation is not without its problems. With the teacher as the focus, use a radio-microphone on her, with a receiver plugged into the camera to simultaneously record with the picture. The teacher's voice is clear, but students' more variable, especially quiet-spoken girls. The camera microphone is almost useless, except for teacher talk to the whole class. The researcher also needs to try to keep notes on the teacher-student interactions as it is not always possible to be sure, when the video-recording is viewed, who is talking in a pair of students. If a radio-microphone is not available then second best is to give the teacher a small portable cassette recorder (or digital equivalent) to put in his pocket (or slung in a passport wallet if she has no pockets). The problem here is synchronising the visual and audio records.

If the students are the focus, then both the visual and audio record are problematic! If the students remain at their table or bench, the camera can be focused on them as a pair; it can be placed on a tripod if what they are doing is always visible (or less important than say the talk). As soon as one or both moves, there is a decision as to who to follow! There is no simple guidance, and whichever one is not videoed, notes will have to be made on the other one. If the group size being observed is larger than two, then the problem is too!

The audio recording is made problematic because a camera microphone will not be reliable, a table microphone will continually pick up the noises of their activity (banging and scraping), leaving really radio microphones as the only reliable means. However, for more than one student this implies either mixing the two inputs prior to feeding into the camera, or recording each one separately on a separate audio tape, with the synchronisation problem mentioned earlier. We recorded four students at a time on a standard audio-cassette using a multi-channel system, but it was only possible to transcribe this manually as no transcribing machines could cope with four-track recording. The technology is probably out there, but time will be needed to make sure that all parts of the process can cope with it (i.e. data collection, transcription and analysis). Whatever approach is taken to audio and video recording, it is essential to do some trial recordings in each classroom with the teacher and students involved.

This raises one small point on digital media. The quality is good, but it think through the replay and transcription issues. It is essential to be able to copy tapes (never depend on just one copy), and then to easily replay them. If a professional transcriber is involved (even for interviews), then this person needs to have compatible transcribing equipment (often these days still in analogue form). Viewing video on digital equipment is fine provided there is access when, and for as long, as is needed.

Interviewing

As with all data collection, the main issue is to be clear about the focus. Even in an unstructured interview, there should be a clear idea of what the questions are trying to elucidate. This requires an interview schedule and a commentary on how the interview questions relate to the research questions. As with observations, structured interviews are possible if there are clear questions and the responses can be anticipated; the interviews here seek to gauge the range or extent of opinion on something. Unstructured or semi-structured interviews allow for more context-dependent answers or complex relationships of issues or just to be able to explore an uncertain area.

I have already noted the difficulty with interviewing inarticulate students. One solution is to have a practical focus or to have objects to talk about (e.g. a project file and products designed and made). Figure 3 shows an extract from an interview schedule to study student ideas on electronics. This approach is likely to produce not only more meaningful (valid) data, but more reliable data. It is also possible to show students extracts from a video of their lesson to ask them about the activity (stimulated recall), but this requires time to analyse the lessons first, then to be able to go back to the school and show them to students (see Murphy, Issroff & Scanlon, 1995). More commonly it is possible to discuss the lesson video with the teacher and, for this, it might be enough to give her the tape to view, though of course this might be an uncomfortable experience for her (e.g. Murphy & Spence, 2001).

1a What do we mean by electronics? What comes to mind when you think of the word 'electronics'?

1b What does electronics do?

1c *Ask questions around example objects*

(i) Does this involve electronics? (*examples to show: watch; mobile phone ...*)

(ii) Where's the electronics in this? (*examples to show: IC board (PCB); PIC chip in a board; 'raw components' in PCB*)

2. Have you enjoyed this project that you have been doing that uses electronics?
What made it particularly enjoyable or not enjoyable?

3. Are projects that use electronics more or less interesting than other D&T? Why is this? *[If student says it depends on which D&T subject then ask them to compare it with one that uses the same form of materials e.g. resistant, or textiles etc.]* ...

Relevance: Think about any projects in D&T that really got you interested. What was it about them that made them especially interesting? How does your recent electronics project fit in with that?.....

Figure 3: An extract from a student interview schedule on electronics

Finding time to interview students (and indeed teachers) is a problem and it is best done within the lesson. Often technology lessons have so much individual activity that asking a few questions of students for 5-10 minutes is feasible. Teacher interviews invariably have to be conducted outside lessons and, if the focus needs to be a particular lesson, then this often results in a hasty conversation immediately after the lesson. However, most teachers interview well and can talk about their general views on technology, how to teach and learn it, and about particular projects and students. They are less able to recall specific events in the classroom, unless it is fresh in their memory.

Exploring learning outcomes

Any assessment information that is available for students can be used, though it seldom relates to the focus of research. Even if it is the grade for a particular project, this only gives gross information, unless there is a breakdown of marks, such as is given in public examinations with elaborate marking schemes. Constructing tests is a skilled and time consuming process (as Kimbell, 1997, amply testifies). In a case study it is possible to produce outcome information that can be valid, though it may not have measured reliability. But it is possible to ask students assessment-like questions to gauge their understanding on, for example, their prior science knowledge or systems ideas.

In our study of science knowledge in a technology project (Levinson, Murphy & McCormick, 1997, p. 239-241), we asked students:

1. whether they had previous experience of electrical circuits;

2. to make a bulb light, unaided, using simple electrical components, wires, bulbs, switches, batteries, etc.;

3. to make the bulbs shine more brightly;

4. to identify electrical conductors from a sample of objects (metal paper clips, plastic pens, aluminium foil etc.)....

Similarly, to test their ideas on systems we questioned them about an electronic badge with a light sensor, and a moisture sensor, they are working on. They were asked to identify the system 'input', 'process' and 'output', which resulted in some interesting answers, especially about 'process' (McCormick & Murphy, 1994; Levinson, Murphy & McCormick, 1997, p. 243).

ANALYSIS OF VIDEO

This is the most problematic part of the analysis for a technology education classroom researcher. The recording difficulties are not insignificant, though inexperienced researchers often resort to this approach because it seems to ensure that all the data are collected. That this is not the case should now be abundantly clear! After recording, the problems of analysis remain, and the main issues are data selection and reduction. I have already shown some of the approaches to produce transcripts and lesson analyses, which will maximise quality (reliability and internal validity), so here I will discuss missing steps and additional ideas, especially related to researching knowledge issues.

The most difficult process is to become completely familiar with the data so that important events, connections and possible explanations become evident.

Transcribing video recording of lessons or audio interviews is one way of doing this and it is helpful to keep an analytic diary of ideas and issues that occur (along with appropriate links to evidence), to return to during later analysis. In our studies of problem solving we initially adopted a tool borrowed from systematic analysis, putting the transcription (as in Figure 1) onto a schedule. Table 1 showed an extract of the D&T project that categorises the actions and interactions into six codes representing: teacher's task structuring (TT), teacher-initiated to class (TC), teacher-initiated to a single student or small group (TP), student-initiated to teacher (PT), student to student (PP), and student actions (P). There are a number of detailed codes indicated at the beginning of an action or interaction (e.g. PI for 'student insight'), along with the students' names (T & K) and the teacher's name (RW). Alongside this are contextual statements or (in italics) an analytic or subjective comment, for example, puzzles in the behaviour (e.g. *'Why has she not copied this?'*). The first intention was to use this in some systematic way, but it soon became clear that, while this detailed coding helped us to carry out a subsequent analysis, we could not use the codes in any quantitative way (e.g. by totalling up for a lesson or across lessons). The process led us to spend a lot of time thinking about the behaviour, and enabled us to isolate critical incidents either for problem solving or knowledge issues. We did not define a critical incident in any formal sense, but built our analysis around them. We subsequently drew up other tables that listed situations, student problems, evidence for knowing this was a problem, the consequences of the problem for the student and the teacher, and the outcome of the situation (e.g. solution, or otherwise, of the problem). Table 1 shows an analysis of an early part of a lesson showing the use of concepts or ideas by both teacher and students in the electronics badge project, discussed earlier.

In the end these various kinds of analyses lead to narratives about the processes of problem solving or of how knowledge issues are experienced and dealt with, and only rarely is there any quantitative analysis. The narratives are structured around types of incidents that are based on analytic categories such as 'knowledge on a need to know basis', 'the issue of transfer', 'the link of procedural and conceptual knowledge' and 'knowledge for action'.

Table 1: Concept and terms used in the first session of an electronics project

Time	Concept or idea	Teacher's term or analogy	Students' term or idea (prior knowledge)
10:15	Circuit	Running track	D: goes in a complete circle ... not a gap in it J: not broken
10:16	Battery	(accepts B's term) Provides the force	B: power source
10:16	Battery positive and negative	Starts at one point	
	Battery symbol and polarity	[visual] Cell is part of battery	D says it is 'a cell'
10:15 - 10:16	Direction of flow (what 'flows' is unspecified)	It goes from ... and goes to In fact he says this is not what happens [implying electron flow in opposite direction?]	D: from positive sign .... to negative sign
10:17	Flow of electrons around circuit		
10:17	Control of flow of electrons	Implies a resistor by accepting D's response	Switch or a resistor
10:17	Resistor	Slows [electrons] down	Lots of little wires ....it has to go further
	Volts	'v' as battery voltage	Volts
	Simple resistor	Restrict and direct flow of electrons	
	Light dependent resistor	[visual] LDR	Solar panel
	resistor colour codes (being able to interpret)	Four coloured bands give value: first number, second number, number of zero's	[new knowledge]
	resistor tolerance (final colour band)	Amount [of resistance] above and below	[new knowledge]

'B', 'D' and 'J' are students that could be identified during the interaction.

STRENGTHS AND WEAKNESS OF THE APPROACH

In the end the quality of technology classrooms will determine the quality of students' technology education. That being the case, we need to use research to understand what goes on in them. Such classrooms are complex, and the case study is ideal for representing such complexity and how the various activities and interactions lead to particular outcomes. Further, it does this in ways that teachers can recognise as authentic, and which carry general lessons for them and their students. They represent real classrooms and, if the evidence is used to justify and explain the general issues that are derived from the case studies, this can lead to both strong theoretical generalisations and professional learning on the part of teachers. There are a growing number of such case studies and these generalisations are increasing and improving. Where a research topic is in its early stages, as with, say, qualitative knowledge in the technology classroom (e.g. McCormick, 1999), case studies are ideal for exploring the issues, and in laying the ground work for further case studies that can be more explanatory.

But there is a down side to these kinds of studies. Most important, they require a great deal of skill in formulating the research questions, conceptualising the issues involved, in collecting the data, and in carrying out an analysis and writing the case study itself. The work is very intensive, in its use of research resources, and in the work to be carried out. For example, 30-40 hours of video for a single case study of a project, with a number of interviews with the teacher and one or two with say four students, along with all the associated documents, is a lot of data to work with. For a single researcher it is demanding of analytic powers and of persistence to transcribe the actions and talk faithfully. Case studies can also be risky to undertake. If time and effort is invested in setting up the study and in choosing teachers and school situations, but then it turns out not to be as fruitful as first thought, a lot of effort and research resource may have been wasted.

Where the theory is weak or not articulated, or where there is not much previous work to build on, such case studies may be too descriptive to give any analytic purchase and have little significance to those outside the immediate circle in the case situation. If the purpose of the study is to see what happens in some new situation or where we know nothing, this description might be important. One of the failings of classroom case study research, however, is the lack of replication of previous studies, a failing actually of much of educational research (Hargreaves, 1996). So much has yet to be done in this field that researchers tend to work on a wide front, and from different perspectives, with a lack of build up of a body of knowledge.

NOTES

1 I would like to acknowledge the work of my colleagues Patricia Murphy, Sara Hennessy, Marian Davidson, Hilary Evens and Stephen Lunn, all of whom contributed to the development of the methodology and methods related to these various studies.
2 This is explored in Levinson, Murphy & McCormick (1997).
3 Yin (2003) considers other approaches, namely, survey and experimental studies, the latter being an artificial situation constructed so that behaviour can be manipulated to allow the relationships among particular variables to be examined.
4 Yin (2003) gives a fifth reason for undertaking case studies in evaluation research, namely to carry out a meta evaluation (an evaluation of evaluations), but this is a specialised form.
5 Appealing to 'grounded theory' (Glaser & Strauss, 1967) does not preclude acknowledging or building upon what is already known.
6 This table was originally presented in McCormick & Murphy (1994) and subsequently published in McCormick (2006, p. 40).
7 Murphy *et al.* (u.d. a) over-sampled their cases (i.e. chose more than were necessary) because they could not pre-sample suitable teachers for observation.

REFERENCES

Assessment of Performance Unit [APU] (1991) *The assessment of performance in design and technology*. London: School Examinations and Assessment Council.
Becker, H. (1958) Problems of inference and proof in participant observation. *American Sociological Review, 23*, 652-660.
Cobb, P., Confrey, J., diSessa, A., Lehrer, R. & Schauble, L. (2003) Design experiments in educational research. *Educational Researcher, 32*(1), 9-13.
Evens, H., and McCormick, R. (1997) *Mathematics by design: An investigation at Key Stage 3* (Final Report for the Design Council). Milton Keynes: School of Education, The Open University.
Glaser, B. & Strauss, A. (1967) *The discovery of grounded theory*. London: Weidenfeld & Nicolson.
Hammersley, M. (1990) *Classroom ethnography. Empirical and methodological essays*. Milton Keynes, The Open University Press.
Hargreaves, D. (1996) *Teaching as a research-based profession: Possibilities and prospects*, Teacher Training Agency Annual Lecture, London, Teacher Training Agency.
Kemmis, S. (1993) Action research. In M. Hammersley (ed) *Educational research: Current issues* (pp. 171-190). London: Paul Chapman Publishing Ltd.
Kimbell, R. (1997) *Assessing technology: International trends in curriculum and assessment*. Buckingham: Open University Press.
Kimbell, R., Stables, K. & Green, R. (1994) *Understanding technological approaches. Final report to ESRC*. London: TERU, Goldsmiths.
Lawson, B. (1990) *How designers think* (2nd Edition), Butterworth Architecture, Oxford.
Levinson, R., Murphy, P. & McCormick, R. (1997) Science and technology concepts in a design and technology project: a pilot study. *Research in Science and Technological Education, 15*(2), 235-255.
Logan, C. D. (2005) *The representation of knowledge and expertise in the undergraduate graphic design curriculum*. Unpublished Ed. D. thesis, Milton Keynes, The Open University.
McCormick, R. (1997) Conceptual and procedural knowledge. *International Journal of Technology and Design Education 7*(1-2), 141-159.
McCormick, R. (1999) Practical knowledge: A view from the snooker table. In R. McCormick & C. Paechter (Eds.) *Learning and knowledge* (pp. 112-135). London: Paul Chapman.
McCormick, R. (2006) Technology and knowledge: Contributions from learning theories. In J. Dakers (Ed.) *Defining technological literacy: Towards an epistemological framework* (pp.31-47). London: Palgrave.
McCormick, R. & Davidson, M. (1996) Problem solving and the tyranny of product outcomes. *Journal of Design and Technology Education, 1*(3), 230-241.
McCormick, R. and Murphy, P. (1994) *Learning the processes in technology*. Paper presented at the British Educational Research Association Annual Conference, Oxford University, England, September.
McCormick, R., Murphy, P., Hennessy, S. & Davidson, M. (1996) *Problem solving in science and technology education*. Paper presented at symposium 'Teaching and learning procedural knowledge in science and technology', Annual Meeting of AERA 1996, New York.
Murphy, P. (2006) Gender and Technology: Gender mediation in school knowledge construction. In J. Dakers (ed.) *Defining technological literacy: Towards an epistemological framework* (pp. 219-237). London: Palgrave.
Murphy, P., Davidson, M., Qualter, A., Simon, S. & Watt, D. (u.d.) *Effective practice in primary science. A report of an exploratory study funded by the Nuffield Curriculum Projects Centre*. Milton Keynes: Centre for Curriculum and Teaching Studies, The Open University.
Murphy, P., Issroff, K. & Scanlon, E. with Hodgson, B. and Whitelegg, E. (1995) Group work in Primary Science - emerging issues for learning and teaching. In A. M. Anderson, K. Schnack & H.

Sørensen (Eds.) *Science –natur/teknik, assessment and learning. Studies in Educational Theory and Curriculum*, Volume 22 (pp. 47-69). Copenhagen: Royal Danish School of Educational Studies.

Murphy, P., McCormick, R., Lunn, S., Davidson, M. & Jones, H. (2004) *EiS Final Evaluation Report. Evaluation of the promotion of electronics in schools regional pilot: Final report of the evaluation.* Milton Keynes: The Open University.

Murphy, P. & Spence, M. (2001) Teacher knowledge and curriculum reform. *Curriculum and Teaching Dialogue*, 3(1), 39-45.

Shayer, M. (1999) Cognitive acceleration through science education II: its effects and scope. *International Journal of Science Education 21*(8), 883-902.

Stables, K. & Kimbell, R. (2006) Unorthodox methodologies: Approaches to understanding design and technology. In J. Dakers (ed.) *Defining technological literacy: Towards an epistemological framework* (pp. 313-330). London: Palgrave.

Stenhouse, L. (1982) The conduct, analysis and reporting of case study in educational research and evaluation. In R. McCormick *et al.* (Eds.) *Calling education to account*. London: Heinemann Educational Books/The Open University Press.

Walker, R. (1986) The conduct of educational case studies: Ethics, theory and procedures. In M. Hammersley (Ed) *Controversies in classroom research* (pp. 187-219). Milton Keynes: The Open University.

Yin, R. K. (2003) *Case study research: Design and methods*. Thousand Oaks, California: Sage Publications.

Robert McCormick
Centre for Research in Education and Educational Technology,
The Open University
United Kingdom

FRANK BANKS

INTERNATIONAL COLLABORATIVE CASE STUDIES

Developing Professional Thinking for Technology Teachers

INTRODUCTION

This chapter describes the stages of a project that used multi-site case study methodology for an international collaboration project – Developing Professional Thinking for Technology Teachers (DEPTH). Following preliminary work to establish and then test a possible common framework for analysis, the DEPTH project was a study conducted with both primary and secondary technology pre-service teacher education students in countries as diverse as Canada, Finland, New Zealand and the United Kingdom who were given the same teacher-knowledge graphical framework as a tool to support reflection on their professional knowledge. As we will see, despite the different country contexts, student teachers of technology could use the framework to articulate aspects of their developing teacher knowledge. The common graphical tool enabled them to set out their subject knowledge, pedagogical knowledge and 'school' knowledge and was useful in helping them become more self-aware. The journey towards the DEPTH collaborative multi-site case studies was from a small single site case study which proposed, originally rather tentatively, an analytical and theoretical standpoint, then on to a two-site study to test the model and finally to the collaboration between researchers in different country contexts.

THE EARLY BEGINNING

As McCormick explains in Chapter 2, case study methods are often used in situations where the context is rich and which has a bearing on the phenomena that is being investigated. The series of linked case studies that became the DEPTH project has its beginning in a small scale study to try to investigate the teacher knowledge of novice technology teachers and how that developed during initial teacher training. At that stage there was no thought of a wider study. As we will see later, the ability to collaborate on such a project as DEPTH, without external funding, was a feature of the relevance of the topic to all teachers developing their understanding of the relatively new curriculum area of Technology, and the almost 'pioneering spirit' of researchers in school technology who meet up at meetings such as the PATT (Pupils Attitude Towards Technology) series of conferences.

The motivation for the original small scale study was threefold:
- the need to find a structure and 'vocabulary' to facilitate conversations between student teachers and their school-based mentors following a move in England to

place student technology teachers into school for a longer period of time than had previously been the case.
- the desire to explore the concept of pedagogical content knowledge, which was gaining prominence in the early 1990s, in the relatively new subject of technology
- the desire to see if the different Design and Technology (D&T) teachers viewed their knowledge in different ways and how their view of the subject impacted on what they considered appropriate teaching strategies.

Current teachers of technology often complain that graduate student teachers have the depth but not the breadth of subject knowledge now required to teach the school subject adequately. But this possible mismatch of subject content knowledge is not the only issue. In a school's technology faculty there is a collection of different pedagogic styles developed from the several ancestors in manual training. The students, therefore, will also have different experiences and expectations of their teaching role in the classroom. Their developing 'teacher professional knowledge' draws firstly on their subject content knowledge, but also on their embryonic knowledge of such matters as the curriculum materials available, their understanding of what is considered important for the subject as it exists in school technology and pedagogic content knowledge. The first case study tried to explore that developing professional knowledge.

THE FIRST CASE STUDY- THEORY DEVELOPMENT

Nine student technology teachers were followed through the eighteen months of the UK Open University's (OU) Technology Postgraduate Certificate in Education (PGCE) course (Banks, 1996). The students were interviewed early in the course shortly after they had experienced three weeks school experience of a mainly observational nature. The interviews followed a semi-structured format in order to explore those aspects of their personal histories which the student considered would be most useful to them as a teacher but were quite wide ranging, in the hope that this would allow the students to make explicit the taken-for-granted assumptions about teaching and children which they held. All interviews took place in the student's home.

Five of the students were visited on subsequent school experiences and their mentors interviewed, school documentation collected, and the context in which the student was working observed. This observation confirmed the range of different curriculum structures which schools had established to teach the national curriculum and consequently the different expectations of the kind and level of knowledge: subject, pedagogical and curricular which mentors expected of the student.

Finally, at the end of their last eight-week teaching placement, seven of the students were interviewed again. The interviews were transcribed and analysed and it became clear that there were different aspects of teacher professional knowledge that interacted in any classroom context. The following vignette from this initial study demonstrates this.

ROBERT MCCORMICK

Exploring teacher professional knowledge: A case study vignette

Two students were observed in December 1995. Martin and Mark had planned and begun to pair-teach a series of lessons for their placement school. The technology department was concerned that the existing school scheme of work which was offered in year 7 (11 year olds) did not yet include aspects of simple electronics. The mentor asked Martin and Mark, working as a pair, to organise the teaching of this. The mentor herself lacked subject knowledge in this area (having a business studies background) and asked the students to come up with the resources for a project which the whole department could use.

Although some advice was given by the Science department (who were concerned that the pupils should not be confused by teaching carried on elsewhere in the school!), the students were largely left to themselves. Using their own ideas and curriculum materials such as text books and electronics kits already in the school, the students decided to organise their teaching around the development of a "face mask with flashing eyes". They found this a very difficult exercise. A particular lesson concerned the pupils investigating which materials were conductors and which insulators. To do this the student teachers employed a standard kit called 'locktronics' but talked about the circuit by drawing diagrams on the chalkboard.

The purpose of the project was unclear in the minds of the student teachers. The face mask was considered primarily as a means to teach aspects of electronics and the functional aspects of wearing the mask were not thought through. For example, the students had not considered where the battery would be located on the mask or how it would be supported. They thought that an understanding of V=IR was important, but the science department had suggested that the use of such an equation was too difficult for 11 year old pupils. Their desire to teach the *science* subject background, such as (in this lesson) conductors and insulators and the existence of electrons, cut down on the time for making. They were unclear if the overall purpose of the activity was designing, acquiring specific skills, or a "seeing-is-believing" confirmation of scientific principles.

Only Mark had used the electronics kits before as a pupil, and both student teachers were unfamiliar with the way they could be used in the classroom. The pupils had some difficulty in manipulating the components and interpreting the circuits which they had constructed on the boards. For example, the pupils did not easily link up the connectors to make the bulb light as they invariably first constructed a loop of wire to the bulb before connecting the power supply (referred to as a battery in the original explanation by the student teacher). Later the pupils did not see how the kit could be adapted to accommodate different shaped rods of various materials.

The students own understanding of simple electricity was sufficient, but lacked the "flexible and sophisticated" (McDiarmid *et al.* 1989) features to ensure that it was conveyed clearly. For example, a description of current flow also involved a confusing discussion of electron flow. A picture of a battery was combined (incorrectly) with a diagram of the electrical symbols. The rather unsatisfactory chalk-board illustration shown in figure 1 was the result:

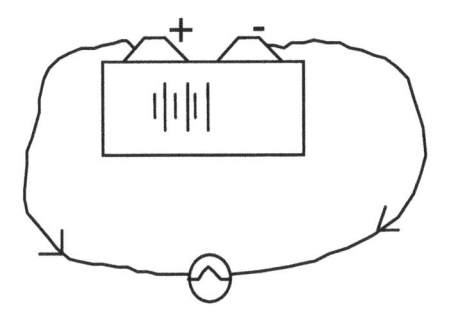

Figure 1: The final chalkboard diagram confusing current and electron flow

As these students were not able to enlist the experience of their mentor, they drew on their own embryonic pedagogical content knowledge (see below) to formulate teaching activities for the project. They naturally used analogies to try to convey ideas about electrical flow. For example, Mark talked about how it is easier to walk around a hill, rather than walk over it, in an attempt to cover the idea of a "short circuit" quickly. As they considered an understanding of electrons an essential pre-requisite to an understanding of 'conductors and insulators', Martin showed a real model Figure 2 and talked about it using the chalk-board diagram Figure 1.

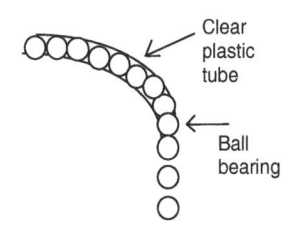

Figure 2: A model to illustrate electrons in a wire

The tube, shown to the pupils later, represented the wire and the ball bearings were the electrons. It is unclear what the pupils thought about the size of electrons and the need for a conductor for electrical flow! The student teachers wished to scaffold the learning of the pupils and that they believed a "hands-on" approach was appropriate. However, they found it difficult to leave the pupils to experiment with the kits, and continually intervened to move them on as time was felt to be so short. Too much was attempted too quickly and some pupils became confused then bored. The students did not have the experience to know which aspects of electricity are difficult to convey. Indeed, they were unsure of how all this fitted into "school knowledge" as they were unclear about why they were teaching this knowledge in relation to this particular design-and-make project.

Theory development

The vignette above from the first study illustrates what was being observed and through interviews with Martin and Mark and other students, key aspects of teacher professional knowledge became apparent. At the same time the literature on teacher development was explored.

Since the mid-1980s there has been considerable discussion and a growing body of research on the forms of knowledge required by teachers in performing their role (Shulman and Sykes, 1986; Shulman, 1986; Grossman, Wilson and Shulman, 1989, McNamara, 1991).

Shulman, for example, asks the central question:

> How does the successful college student transform his or her expertise in the subject matter into a form that high school students can comprehend? (Shulman 1986)

The different forms of teacher knowledge have been usefully summarised by McNamara (1991, p 115), adapted here:

Subject Content Knowledge
- If the aim of teaching is to enhance children's understanding then teachers themselves must have a flexible and sophisticated understanding of subject matter knowledge in order to achieve this purpose in the classroom.
- The understanding of subject must be "flexible and sophisticated" to include the ways in which the subject is conducted by academics within the field, "to draw relationships within the subject as well as across disciplinary fields and to make connections to the world outside school" (McDiarmid et al 1989, p193)
- Teachers' subject matter knowledge influences the way in which they teach, and teachers who know more about a subject will be more interesting and adventurous in their methods and, consequently, more effective. Teachers with only a limited knowledge of a subject may avoid teaching difficult or complex aspects of it and teach in a manner which avoids pupil participation and questioning and which fails to draw upon children's experience.

Pedagogical Content Knowledge
– At the heart of teaching is the notion of forms of representation and to a significant degree teaching entails knowing about and understanding ways of representing and formulating subject matter so that it can be understood by children. This in turn requires teachers to have a sophisticated understanding of a subject and its interaction with other subjects.

Shulman states:

> Within the category of pedagogical content knowledge I include, for the most regularly taught topics in one's subject area, the most useful forms of representation of those ideas, the most powerful analogies, illustrations, examples, explanations and demonstrations - in a word, the ways of representing and formulating the subject that makes it comprehensible to others. (Shulman, 1986)

Paramount amongst school subjects, technology is characterised by a pedagogy where there is no 'right answer' but rather different responses to the same problem are valued - some judged better than others. Compared with other subjects such as science and mathematics, perhaps a teacher of technology is less in a position of being a 'fount of all wisdom' as might be suggested by the Shulman's quote above, but rather a guide to help a pupil to (as Barlex would put it) 'Design what they Make and Make what they have Designed'. This is not to deny the important role for subject knowledge in Technology, nor to suggest that the teacher is not an important source of information, but the teacher's knowledge and expertise need not be a brake on the speed or direction of the pupils' development or creativity. For example, in electronics a pupil can treat an amplifier as a 'system element' without knowing or needing to know the details of the physics of its operation. Similarly a pupil can make artefacts using a polymer without needing to know much more than the underlying concept of giant molecules and the interaction between the chains. However, that pupil may indeed need to know more sophisticated ideas about amplification or plastics as their interest in their design problems develop. A teacher who is able to engage them in a conversation at an appropriate level will be better able to match the curriculum to the pupil.

In contrast to Shulman, Gardner's (1983) work is rooted in a fundamental reconceptualisation of knowledge and intelligence. His theory of multiple intelligences allows us to view pedagogy from a perspective on student understanding. In common with the ethos of School Technology, the focus shifts from teacher to learner, from technique to purpose. The five "entry points" which Gardner proposes for approaching any key concept - narrational, logical-quantitative, foundational, experiential and aesthetic - do not simply represent a rich and varied way of mediating a subject. Rather they emphasise the *process* of pedagogy and a practice which seeks to promote the "highest level of understanding possible" (Gardner 1991). This line of thinking persuaded us that a wider term 'pedagogical knowledge' was more appropriate.

School-Subject Knowledge

To these types of teacher knowledge it became clear that we would wish to add 'school-subject knowledge'. By altering technology to make it accessible to learners, a distinctive type of knowledge is formulated in its own right - 'school technology'. In the same way that school science has differences from science conducted outside the school laboratory, so school technology is different from technology as practised in the world outside the school. As a "subject designed by committee", the school knowledge of Design and Technology is particularly specific and rarely exists as a coherent body of knowledge outside the classroom. But the subset of technological knowledge which is 'school technology' is a function of the schooling process and so would exist even without a national curriculum to guide its formulation.

La transposition didactique Chevellard (1991) defines as a process of change, alteration and restructuring which the subject matter must undergo if it is to become teachable and accessible to novices or children. This has echoes in Shulman's 'pedagogical content knowledge' but this builds on work done by Verret (1975). Verret's original thesis was that 'school knowledge', in the way it grows out of any general body of knowledge, is inevitably codified, partial, formalised and ritualised. Learning in that context is assumed to be programmable, defined in the form of a text, syllabus or national curriculum, with a conception of learning that implies a beginning and an end, an initial state and a final state. Verret argues that knowledge in general can rarely be sequenced in the same way as school knowledge and that generally, learning is far from being linear. Our thinking developed this to appreciate that school knowledge is greatly informed by the local school ethos, common practices and the authenticity of the activities that pupils are required to undertake.

These different categories of teacher knowledge for technology teachers are summarised by Figure 3 (Banks, Leach & Moon, 1999). The diagram has some similarities with the developmental model of 'pedagogical content knowing' proposed by Cochran, DeRutter and King (1993), but is simpler in form. The diagram tries to indicate the *synthesis* of these types of teacher knowledge and the inadequacy of the picture to do so is recognised. One might initially see 'school knowledge' as being intermediary between subject content knowledge (knowledge of technology as practiced by different types of technologists) and pedagogical content knowledge as used by teachers ("the most powerful analogies, illustrations, examples, explanations and demonstrations"). This would be to underplay the dynamic relationship between the categories of knowledge implied by the diagram. For example, a teacher's subject knowledge is enhanced by their own pedagogy in practice and by the resources which form part of their curricular knowledge. What teacher has not confessed to only really understanding a topic when they had to teach it to others! All these types of teacher professional knowledge are strongly influenced by the personal subject construct of the teacher.

Personal Subject Construct
The past experience of learning technology, a personal view of what constitutes "good" teaching and a belief in the purpose of technology for all underpins a teacher's professional knowledge. This is as true for any teacher. A student teacher has to question his or her personal beliefs about their subject as they work out a rationale for their classroom behaviours. But so must those teachers who, although more experienced, have undergone profound changes of curriculum emphasis during their career.

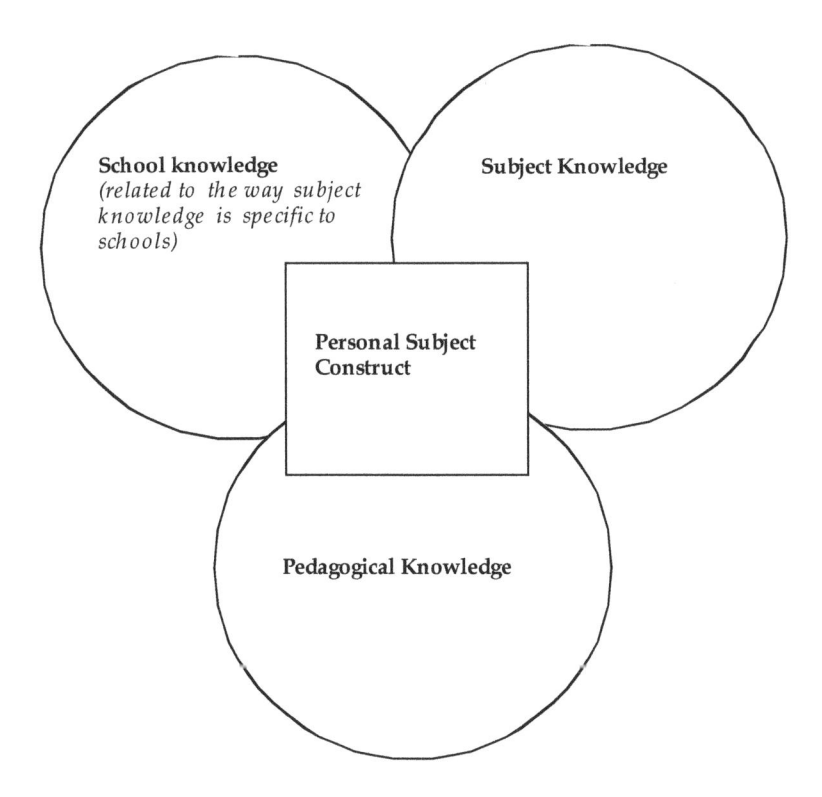

Figure 3: Aspects of teacher professional knowledge

THE MOVE TO A TWO-SITE CASE STUDY

Although the model suggested by the first case served its purpose of providing a framework and vocabulary for student teachers, school-based mentors and college lecturers to share, was it valid? External validity in case study research is an indication of the extent to which a phenomenon believed to be observed in one instance is applicable in other cases (See Yin, 2003, p34 and McCormick Chapter 2 in this volume). In particular, was the model itself useful (Fig. 3) meaningful to teachers in different contexts? In some respects this external validity had been shown as student teachers and mentors from different schools had contributed to its development, however, they all shared the same Open University course context. To test the external validity of the case, a collaborative project was set up between two institutions in the United Kingdom (Banks & Barlex, 1999). This time the graphical tool, used as a means of analysis in the first study was used instead as a research instrument in the two-case study.

Thirteen technology student teachers in the final year of their course from Brunel University and the Open University were interviewed and shown a blank outline of the Teacher Knowledge Framework (Fig 3). The different elements of the framework were explained to them and, in relation to the work on teaching placement, they were asked the following:
- What subject knowledge (about D&T) do I have/need to get for the teaching?
- What pedagogic knowledge (about teaching methods) do I have/need to get for the teaching?
- What school knowledge (about ethos, procedures, significance of some activities) do I have/need to get for the teaching?

The students were also asked to consider their 'personal subject construct' as outlined above.

As might be expected, the students used the framework with a range of levels of sophistication. However, for all students it provided a meaningful and useful focus for debate as is shown by some short extracts from the many students comments under the different categories:

School knowledge

Frank: After a few weeks within the department I noticed that the department ethos, or approach to teaching was the same across the board. [...] The projects from year 7 upward were very closed in nature and pupils were led by the hand through each assignment. This resulted in the pupils producing an end product identical to everyone else. I must admit it was to a high standard and I learned a lot about subject knowledge, especially in the area of woodwork practices and processes. It seemed to me that the department was setting Design and Make assignments that were in fact Focused Practical Tasks.

Christopher: In school you have to work in a particular way. For example the control software package configures the way I have to work and the way pupils have to think because that is recommended by the exam board.

Vincent: In this school the department is driven by the exam. That is all that is important. So I think technology here is too individualistic where industry is social.

Subject knowledge

All the students could identify subject knowledge gaps that they had. Indeed the rectification of technology subject gaps is a pre-occupation on many teacher preparation courses at all levels (see Banks 1997):

Colin: There is no 'big hole' in my knowledge due to being a technician but sometimes I forget the 'easy stuff'!

Frank: I knew I was lacking in some of the graphics subject knowledge, so I spent a great deal of time and effort getting up to scratch on them.

Pedagogical knowledge

Frank had a clear view of how the pupils' enthusiasm for technology and the quality of their work was intimately bound up with the teaching strategies deployed.

Frank: During my last school practice I worked with a Year 10 Graphics group. I found this group to be very passive and generally switched off to the subject. [...]. There was a lack of imagination being demonstrated in their work, which I felt, was coming from the way the subject was being presented to them. [...] I introduced group-work, which they had not experienced in D&T and let them give their opinions. We explored ways we could use these skills in presenting what we wanted to say in graphics. [...] The pupils responded very well and produced many varied and imaginative results. In addition [...], I set a competition to produce a graphic image. This was very open, with the only criteria being that it was interesting to the eye or not as the case may be. This really was difficult for the pupils to take on board initially, as they wanted to know what I wanted as a result. In the end they produced a very good series of images, some 3D, some computerised, some with alternative backgrounds. The approach to teaching in a different way from chalk and talk seemed to awaken this group of pupils.

Personal subject construct

The nature and quality of the answers showed the range of personal subject constructs held by the student teachers and they often mentioned how it conflicted with the construct held by their school mentor or other people in the school technology department. Vincent saw technology as being closely linked to 'real-life' and vocational preparation. Christopher, in contrast, saw technology as empowering for the pupils. They should understand 'how to wire a plug and not be scared to do things'. He wanted pupils to 'have a go'.

Although the extent of this two-site case study is still limited in the number of students who took part, it is significant in that the students across two quite separate institutions, with students from very different parts of the UK, could identify with the concepts outlined in the Teacher Knowledge Framework. It is clear from the above extracts and examples that they could use the categories as a means to reflect on their practice. The investigators could, in turn, use the diagram as a way to group aspects of teacher knowledge when the students described both their own practice and that of their colleagues in school.

THE MULTI-SITE CASE STUDY: DEPTH

Setting up a set of multi-site international case studies is a formidable challenge. However, DEPTH is an example of how the opportunity offered by researchers meeting to discuss their work, the interest generated by a topic which has common appeal and a methodology which can be used in a range of contexts can come together in a successful collaborative project (Banks *et al*, 2004).

Following the same structure as the two-site case study at Brunel and the Open University, the research was extended into a multi-site case with student teachers in four different countries. The study was extended in the UK to the University of Surrey Roehampton. Students in Finland were at the University of Oulu, and the student teachers in New Zealand studied at Massey University. The students at Queen's University in Canada were interviewed by staff at Brunel University. Following the experience of the two-site study, the students were given an introduction to the Teacher Development Framework (Figure 3), but then the way in which the responses were required were varied according to the context in which the work was taking place. For example, this might be a reflective activity following a long teaching placement, or an on-line activity between remote learners, or an activity following an awareness raising series of tasks. What had originally been used in the first study as a way of summarising observation data of the way teachers were describing their personal teacher-knowledge was now being re-configured, following the two-site intervention study methodology, to be used as a research tool with student teachers to stimulate self-reflection and to record that developing awareness of their teacher knowledge.

The University of Surrey Roehampton

Prior to starting their second long, final school teaching practice twenty-seven secondary and eighteen primary postgraduate students were introduced to the simple framework tool (figure 3) and asked to read a paper (Banks & Barlex 1999) reporting on the two-site study. They were told that they would use the framework to reflect on their practice when they returned to the university. The same approach was used with a group of primary specialist D&T students in their final year of a four-year degree course.

The session for both groups began with a group discussion by the twenty-seven students based on the three elements. School knowledge was discussed first, followed by pedagogical knowledge and finally subject knowledge. This was a deliberate strategy to prevent them focusing too early on subject knowledge. They were asked to think about:
– what are the important issues?
– how did I deal with them?
– could I have done it another way?

The discussion was wide ranging. After the discussion they were asked to write down their thoughts using the framework which exposed their 'personal subject construct'.

Oulu University

In this study the researcher arranged an interview session with the students who had just finished one of the teaching training periods. They did the training in primary schools with 5-6 grade pupils. He copied an enlarged picture of the framework tool (Figure 3) and explained the idea of the model to the students. During discussion the students wrote their thoughts directly on the enlarged picture of the model. The interviewer also made notes about the discussion on his paper. The teaching topic of all the students' had been similar: model aeroplanes made out of balsa-wood. However, they carried out the teaching placement in different schools.

Massey University

Nineteen 3rd year BEd Teaching students at Massey University College of Education were introduced to the graphical representation (figure 3) and asked the following questions:
– What subject knowledge (about technology education) do I have / need to get to teach?
– What pedagogical knowledge (about teaching and learning) do I have / need to get to teach?
– What school knowledge (about ethos, procedures etc.) do I have / need to get to teach?
– What is my personal subject construct?

These questions were similar to those used in the Open University/Brunel University study. The main difference between the Massey study involved primary specialists, rather than secondary. That said these students had completed a number of papers (courses) in technology education as part of their degree with some completing as many as six.

ROBERT MCCORMICK

Queen's University

Yet another approach to introducing the common framework was set up in Canada. In this study the researcher (a visiting academic) was introduced to a group of fifteen trainee computer studies teachers as an imported robot for which the manual had been lost. He could respond to verbal commands but these had to be precise. A group of students within the group were set the task of giving verbal commands for the robot to make a peanut butter and jam sandwich. This was called the control group. Other groups were given observation tasks with a brief to report back on these observations. One group reported on the subject knowledge used by the control group. One group reported back on the suitability of this approach to teaching programming. One group reported back on organisational issues associated with the exercise and one group was given the task of standing back and reflecting in general terms about the integrity of the exercise. The introduction by the Queen's University lecturer set the scene for giving the robot instructions because he gave "walking, turning and stopping" instructions to get the human 'robot' into the room. The control group found the robot co-operative but limited in that he did exactly what he was told i.e. if told to lift his right arm he'd lift it but keep on lifting or trying to lift it until told to stop. After the sandwich had been made the reports back clearly featured the three parts of the framework tool:
– subject knowledge – if the control group don't give specific instructions the robot fails, just like a computer and they should have taught the robot some sub routines.
– pedagogic knowledge – it would engage children but it took too long. I'd need to find a way of shortening it.
– school knowledge – I couldn't get away with this in my school, the students just wouldn't understand and I wouldn't be able to find a teacher who'd be the robot.

It was only after these comments, however, that the trainee teachers were introduced to the framework (figure 3). They were asked to comment on the session as a whole and to provide written comments as to the value of the subject construct model. Of the twenty-five trainees in the group, fifteen responded.

Results

As was shown by the two-site study, the students used the framework with a range of levels of sophistication. For all students it provided a useful focus for debate, in particular concerning the nature and extent of school knowledge.

School knowledge

    Marko: I was in a remote, countryside field school. There were not very much materials and equipment available…and the tools were not in a very good shape…there was a kind of "laissez-faire" atmosphere in the woodshop classroom…for example, when one of the pupils sawed the [carpenter's] bench, the teacher did not say or do anything…maybe this all is due to the lack of resources. On the other hand, other parts of the school were quite modern. It seemed to be that the teacher who was responsible of the teaching of "technical work" was not very much interested about problem solving approach…

Antti: My training period took place in the city school. There was rather good order in the "technical work" classroom. The teacher seemed to put quite a lot of emphasis on very well finished work…the product/outcome needs to be well done. When I did my training with the pupils they followed instructions of "how to make a wing" [from balsa wood] meticulously….no one made their own solutions.

Student 1 Queen's: The most interesting component of your teaching model is the inclusion of school knowledge – it became clear to me that what and how I was teaching on my teaching round would not have worked for teachers at other schools in Kingston. My school was academically oriented – much more so than others.

Rebecca I was expected to follow exactly what was planned, it makes it difficult to reflect on your own style

Subject knowledge
All the students could identify their subject knowledge gaps:

Emma: I needed to know about smart materials, industrial uses and practices

Mikko: It is essential for teacher to know about the work he is going to present to the pupils. I mean that the teacher should know about the concepts and functional principles which are included in the project or topic. Understanding is important.

Jussi: I had a project where the pupils were making bridges. It is important to know what materials should be used in making different kinds of bridges. Even to me [as a young teacher] it was not entirely clear which materials or structures would be the best. So I had to explore available materials and possible structures before any teaching took place. …. If I had known more about bridges, the pupils could have been able to build more durable bridges with a longer span.

Pedagogical knowledge
The teachers had a clear view of how the pupils' enthusiasm for technology and the quality of their work was intimately bound up with the teaching strategies deployed.

Khan: I felt that the pupils required a change of task setting and a more 3D approach to graphics. [...] I allocated time for the pupils to create prototypes or models from cardboard of the designs they had generated and present their designs to a board of directors (the rest of the class). I left the presentation styles up to the individual pupils but did incorporate some input [...] on presentation techniques using computer graphics and boards. It was expected that the pupils would give a 2-3 min presentation. The end result was a positive change of all the attitudes of the pupils.

Mikko: I think that 'the copying method' traditionally used in educational handicraft should not be used to same extent anymore. If the children are copying the model presented by the teacher, they just learn to copy and maybe some basic skills of using different kinds of tools. But that's all...and the pupils do not get a deeper understanding about the work....

Personal subject construct
When using the model it is easy to focus too closely on the three aspects of 'teacher knowledge' and to direct insufficient attention to the importance of the over-arching influence of the 'personal subject construct'. In the UK this is often associated with the traditional background of teachers prior to the introduction of Design and Technology as a compulsory subject. However, a response from a student teacher at Queen's University was very eloquent in underlining the importance of 'personal subject construct '.

Student 3 Queen's: You asked us to be brutal, so here goes. I think that making a model saying that the three important things about teaching are knowledge of subject, knowledge of pedagogical methods, and knowledge of how to adapt to your particular school, is very wrong. How can you look at teaching and ignore the attitude of the teacher, and their enthusiasm towards their subject, their ability to demonstrate and pass on this enthusiasm, and their interest in and concern for students, among other things? If all that matters in teaching is knowledge, then I think I'm in the wrong profession. But, on the plus side, all three of those things are important to teaching. Or I should say, to students' learning (which is really what it's all about).

LESSONS LEARNED FROM DEPTH

It is significant that intending technology teachers across four counties and three continents could identify with the concepts outlined in the Teacher Professional Knowledge framework (figure 3). It is clear from the above extracts and examples that trainee teachers from both primary and secondary phases could use the categories as a means to reflect on their practice in terms of their current position and where this might lead. The investigators could, in turn, use the diagram as a way to group aspects of teacher knowledge when the students described both their own practice and that of their colleagues in school. So what lessons have been learned? On the positive side:
- The simple framework strikes a chord with both novice and experienced teachers at a range of levels. It has high external validity.
- The framework is not context dependent. The different aspects of teacher professional knowledge can change depending on local factors, such as the

particular country curriculum, the nature of 'school technology' and the rationale for including the subject in the curriculum.
- The different types of 'knowledge' can be interpreted broadly to include processes and skills.
- Many teachers valued the 'process' rather than the result. Whatever the country setting, teachers found it beneficial to focus on their practice and the framework facilitated that endeavour.

However, it has to be recognised that such a liberal interpretation of the meaning associated with different aspects of the framework has its drawbacks:
- Although internally consistent within a country setting, care has to be taken with using the results for comparative purposes. DEPTH is not a comparative education project to, for example, illustrate similarities or differences in curricula. Rather all it can show is that aspects of teacher knowledge are of similar concern in different settings.
- The educational values of the researcher and participants when discussing the frame can influence the way in which the diagram is seen as a 'fixed' entity or a means to promote reflection and discussion. As is clear from Student 3, Queen's comments above, some will reject any attempt to capture aspects of teacher knowledge believing it to be much more to do with ephemeral attitudes and feelings.

DEPTH 2

While recognising the limitations to making generalised conclusions, which is a function of all case study methodology, the perceived usefulness of the framework has resulted in an extension of the DEPTH work in 2005. In this second phase of the project (DEPTH 2) we have developed the line of research in two ways. First, we extended the range of participants to include more experienced teachers involved in in-service work connected to curriculum development. Second, we looked further at the inter-relationship for pre-service teachers between their developing professional knowledge and their own personal subject construct.

The potential for positive impact on pupil learning in technology due to student teachers who are better able to reflect on their practice seems clear from the extracts presented from all the studies. Not only does this framework prove useful in the context of technology teachers in the UK but also for different phase teachers in different parts of the world. As McIntyre (1993) suggests, reflection by novice teachers is very difficult. We believe that this study has shown the framework is a simple yet very effective 'way in' to begin the discussion of the different aspects of teacher knowledge and the part these play in developing a robust personal construct of the subject. Indeed, the discussion of the model itself promotes an insight and often very vigorous discussion into the various aspects that contribute to the professional role of an effective technology teacher wherever they work.

ACKNOWLEDGMENTS

I would like to acknowledge the contribution and insights to the methodology, analysis and theory of the multi-site case study that has been made by my colleagues engaged in the DEPTH project:
David Barlex, Brunel University UK
Jouni Hintikka, University of Oulu, Finland
Esa-Matti Järvinen, University of Oulu, Finland
Arto Karsikas, University of Oulu, Finland
Gary O'Sullivan, Massey University, New Zealand
Gwyneth Owen-Jackson, The Open University, UK
Marion Rutland, Roehampton University, UK
John Williams, Edith Cowan University, Western Australia.

REFERENCES

Banks, F. (1996). *The development of pedagogical content knowledge during initial teacher education* Paper at JISTEC96 Conference, Jerusalem, Israel. January 1996.

Banks, F. (1997). What prior experiences are perceived as useful to students following an ITT design and technology course?, *Journal of Design and Technology Education, 2*(3), 230-235.

Banks, F. and Barlex, D. (1999). No one forgets a good teacher! - What do 'good' technology teachers know? *Journal of Design and Technology Education, 4*(3), 223-229.

Banks, F, Leach, J. and Moon, B. (1999). New understandings of teacher's pedagogic knowledge, in Leach, J. and Moon, B. (eds.) *Learners and pedagogy*, London: Paul Chapman publications.

Banks, F. Barlex , D., Jarvinen, E-M, O'Sullivan, G., Owen-Jackson, Rutland, M.) (2004). DEPTH - Developing professional thinking for technology teachers: An international study. *International Journal of Technology and Design Education, 14*, 141-157.

Chevellard, Y. (1991). *La transposition didactique: Du savoir savant au savoir enseigné*, Paris:La Pensee Sauvage.

Cochran, K.F., DeRuiter, J.A. and King, R.A. (1993). Pedagogical content knowing: An integrative model for teacher preparation, *Journal for Teacher Education*, September-October, *44*(4), 263-272.

Gardner, H. (1983). *Frames of mind: The theory of multiple intelligences*, New York: Basic Books.

Gardner, H. (1991). *The unschooled mind*, New York: Basic Books.

Grossman, P.L., Wilson, S.M., & Shulman, L.S. (1989). Teachers of substance: Subject matter knowledge for teaching, in M.C. Reynolds (ed.) *Knowledge base for the beginning teacher,* Oxford: Pergamon Press

MacNamara, D. (1991). Subject knowledge and its application: Problems and possibilities for teacher educators, *Journal of Education for Teaching, 17*(2), 113-128.

McDiarmid, G., Ball, D.L., & Anderson, C.W. (1989). Why staying one chapter ahead doesn't really work: Subject-specific pedagogy, in M.C. Reynolds (ed.) *Knowledge base for the beginning teacher,* Oxford: Pergamon Press.

Shulman, L.S. (1986). Those who understand: knowledge growth in teaching, *Educational Research Review, 57*(1).

Shulman, L.S., & Sykes, G. (1986). *A national board for teaching? In search of a bold standard. A report for the task force on teaching as a profession.* New York: Carnegie Corporation.

Verret, M. (1975). *Le temps des études*, Paris: Librarie Honoré Champion.

Yin, R.K. (2003). *Case study research: Design and methods*. Thousand Oaks, Calfonia: Sage Publications.

Frank Banks
Centre for Research in Education and Educational Technology
The Open University
United Kingdom

LARS BJÖRKLUND

THE REPERTORY GRID TECHNIQUE

Making Tacit Knowledge Explicit: Assessing Creative Work and Problem Solving Skills

INTRODUCTION

This chapter describes the Repertory Grid Technique (RepGrid), the theories behind it, some illustrative examples and the merits and drawbacks of the method. It will end in a review of different uses of the method and an annotated bibliography of some research studies of relevance.

By describing a not so known method of interview, Repertory Grid Technique, the author shows how to elicit underlying, often tacit criteria that professional teachers use when they assess creative work. It seems plausible that these criteria can be used to enhance a student's development from novice to expert. Some of these criteria may have universal value for the development of skill and creativity in school subjects other than Technology, Art and Craft.

Most subjects in modern schools have an element of creative work. It is a goal of many curricula, to enhance skills and to foster the ability to design and innovate. In Art and Craft Education, this goal plays a major role but it is important in other subjects as well. During the last two decades a new comprehensive subject, "Technology" or "Design and Technology", has been introduced in many countries. In these curricula creative design is a core activity. But, what is creativity? Can it be defined in words? Can it be evaluated? Does any development occur? Can creativity really be taught? These questions are important to address. In the curriculum for the Swedish subject "Teknik" knowledge on these items seems to be taken for granted. The description of the design process is vague:

A practical and inquiry based work will illustrate the design process; Defining the problem, forming an hypothesis, planning, prototyping, testing and modification (Education, 2000, 96).

This is a criticized, simplified, prescriptive and linear description of the design process. Studies of professional designers in action shows that the processes are not linear, they are iterative and very individual. (Mawson, 2003; Middleton, 2005; Petroski, 1996; Williams, 2000)

Traditionally, assessing the finished product, the constructed artefact, the painting, the model of a bridge, has been the way to assess creative work. This will grade the students but will not give any useful clues or feedback to the student as it is focusing on the end result and ignores the process of making. To be able to assess process we need to know more about the strategies, the skills, the abilities, and the habits of mind of experts in the designing task. What behaviour is to be promoted and which signs of progression are to be identified? At first it seems as every design task is unique but studies of experts in different areas show that there are common properties and behaviour to be seen. An interesting study on Art teachers' criteria for creativity and "habits of mind" was done by Lindström et al and is described later on in this chapter. (Lindström, Ulriksson, & Elsner, 1999)

When you study novices becoming experts you will recognize a development and often a change in behaviour during problem solving activities. (Dreyfus & Dreyfus, 1986) The experts seem to be able to concentrate on the salient features of the task, they act fast and proficiently and they share some important habits of mind controlling their design process (Middleton, 2002). Another characteristic of an expert is the inability to verbalize the 'know how' or procedural knowledge because much of it is tacit.

Hillier describes the use of Personal Construct Theory and Kelly's Repertory Grid as a method for making explicit the tacit, implicit, and informal theories that underlie experts practice. Hillier demonstrates that Kelly's Repertory Grid is an appropriate tool for eliciting informal practitioner theory, which is derived from personal constructs and factors. It is a particularly effective method of reflective practice, providing a focus without imposing structures by the interviewer. It provides a rich source of interpretative data, which can be explored collaboratively with the respondents. From this, propositions can be derived which can be tested. These propositions and tenets form the basis of informal practitioner theory. This methodology may provide the means by which formal theories of adult education can be informed by practical knowledge. (Hillier, 1998)

BACKGROUND

Experts and tacit knowledge

In recent years interest in expertise and proficiency has been raising, in educational research, knowledge management and cognitive science.

Stevenson defines expertise as the ability to do something well – better than others just starting out on the undertaking (Stevenson, 2003), He proposes several interesting research questions;

What do we mean by doing something well?

What enables an individual to do something well?

Why does this capacity improve with practice?

Is this capacity confined to a specific field, or is it general?

Can the capacity be learned, and how?

Where is it located?

The quest of eliciting knowledge from experts has been problematic since the beginning of the development of artificial intelligence in the sixties. The database of Expert Systems had to be loaded with knowledge from human experts and these experts seemed to be unwilling or unable to talk about their secrets and methods. When you are using standard interview techniques you are probing the conscious, rational and logical mind of the interviewee. The informant may want to please you and tell you what they think is appropriate, logical and sound. Your data will be full of general rules and standard procedures and not the individuals' own subjective way of coping with problems. His know how or procedural knowledge is hidden even from him because it is tacit.

We know more than we can tell (Polanyi, 1966, 4).

This knowledge is apprehended unconsciously in an implicit way, often outside our own awareness. It is also used in an automatic way and is therefore difficult to elicit by introspection. In cognitive ccience, dual cognitive systems theories have matured during the last 20 years and have given us new ways of understanding tacit knowledge, expertise, intuition, insight and automation. (Cronin, 2004; Epstein, Lipson, Holstein, & Huh, 1992; Ericsson & Charness, 1997; Lieberman, 2000; Nightingale, 1998; Reber, 1989; Sloman, 1996; Sun, Slusarz, & Terry, 2005) Tacit knowledge may, in a very simplified model, be described in the following manner. Individuals store sensory information in implicit memory as signal pattern together with an emotional qualitative assessment of the event. This gives them a tool to make meaning of phenomena in the world just by the recognition of the sensory pattern they experience and what is stored in their implicit library of old experiences. In this way we "learn" what is dangerous and what is not, what is beautiful and what is ugly, what is edible and not. We learn to recognize faces and scenes, sounds and odours. This kind of knowledge might be referred to as patterns of data or information, a "sensogram". There isn't any conscious perception in the classical sense just recognition of similarities and differences with old exemplars. Polanyi (1966), who minted the expression Tacit knowledge, writes about this proximal knowledge that is insignificant by itself, but points to something more important, some distant meaning. The knowledge is contextual and situated as it is apprehended in practice. The professional craftsman learns to feel, listen and smell to decide when the artefact is finished. The surgeon knows how sick tissue feels when she is cutting with the scalpel. The dentist learns how the drill sounds when he reaches fresh pulp. This sensory information is used in several feedback processes and is stored as patterns of sensory information. In the progression from novice to expert this kind of implicit learning is essential (Dreyfus & Dreyfus, 1986).

In the stored "sensogram" there may also be a documentation of internal sensory signals, of how much adrenalin is pumping, which muscles are activated at the moment and other things. This may be used in automatic response- and feedback-system, of the kind that Skinner and Pavlov were describing. Modern experimental psychological research differs from that of the behavioural-movement by allowing memories and emotions into the stimuli-response process.

If we want to elicit this kind of pattern-knowledge we can't use ordinary interview techniques. The information is not stored in a verbal form and the interviewee often don't know it's there and is controlling his decisions and actions.

Not only in artistic judgement but in all our ordinary judgements of the qualities of things, we recognise and describe deviations from a norm very much more clearly than we can describe the norm itself (Schon, 1987, 53).

This is because our ability to recognize patterns and familiarity in an area of our own expertise is strong. We may not always know why but intuitively we feel what is good, bad, beautiful, sloppy, clear, original, etc.

In 1955, Kelly formulated the Personal Construct Theory that tried to explain why people have different views and attitudes towards events in the world. Kelly claimed that during their upbringing people make use of very personal criteria, construct that they use to construe a meaningful world. A construct is not the same as a concept; it is defined as at the same time a similarity and a difference. A construct with one pole described as "a friend" must be described with its contrasting poles and since the opposite of a "friend" could be "foe" but also an "acquaintance" the bipolarity is essential. The opposite of "good" could be "evil" but also "bad" or "nasty".

Kelly designed a method to elicit personal constructs, The Role Construct Repertory Test. The method has been redefined and developed by himself and others and is now known under the name of Repertory Grid Technique (RGT).

RGT has been used in clinical psychology for more than 50 years but since the 1960's has found new use in a variety of research areas. The findings from experimental psychology and cognitive science on implicit learning and knowledge, the ideas of dual cognitive systems and the interest in tacit knowledge have given rise to new expectations for the use of the method. (Gaines & Shaw, 2003) The RGT identifies perceptions, together with associated feelings and intuitions held about the issue in question. Kelly's theory and technique have both been used to explore management and intuitions, which affect behaviour in fields as diverse as quality assurance, performance appraisal, new product development, and consumer choice. (Jankowicz & Hisrich, 1987)

The use of the RGT involves agreement on a topic; the identification or provision of a series of cases, examples, or, in Kelly's terminology, "Elements". It also involves the use of a tightly structured interview in which a systematic comparison of elements enables the respondent to identify "Constructs", i.e., the ways he or she has of making sense of, or construing, the elements. Constructs are frequently expressions of intuitions, "gut feelings," and perceptions, which the individual uses as a guide to action, without necessarily having verbalized them explicitly prior to the interview. There are several software packages that administrate the eliciting process and also supply the researcher with different statistical tools.

However, it is possible to do it manually. The elements are written on separate cards, three cards are selected, tryading, and the subject is asked if two of them share something that separates them from the third element. This property of the two similar elements has to be verbalized into an emergent pole of the construct. The subject is asked to name the opposite of this and this names the implicit or contrast pole of the construct. All elements in the set are rated with this bipolar construct as a yardstick, either to one of the poles, a dichotomy, or on a continuous scale between the poles, a Lickert scale, of 5, 7, 9 or more steps. The result is recorded as an array in the first row in a Repertory Grid. This procedure is repeated until the construct generation is exhausted, or the subject is. It is possible to evaluate small grids manually and some authors think this is essential for understanding the method (Jankowicz, 2004).

The following examples were made using the free version of RepGrid IV that could be downloaded from http://www.repgrid.com. Another free software item consisting of a web based package is WEBGRID III. This can be accessed at: http://tiger.cpsc.ucalgary.ca:1500/WebGrid/WebGrid.html

TO MAKE A REPERTORY GRID INTERVIEW, AN ILLUSTRATIVE EXAMPLE

Research question of the study

The Swedish National curriculum of "Teknik" is goal-oriented, emphazing documented learning outcomes including the development of design abilities and creativity. We ask: What kind of criteria do teachers in the Swedish school use when they are assessing and grading a design and construction project in the subject "Teknik"?

Background

In a comprehensive, lower secondary school in a medium large Swedish city, a class of 15-16 year old students had been working with a project in technology and design. It consisted of two different parts; first an electronic alarm was built following supplied plans and schematics. The teacher delivered the components and the material for the soldering task. The second part of the project was more creative and free. The electronic alarm was to be put into use in a context. This context, a lighthouse, a car, a secret diary etc, was to be designed and built in the form of a model using different kinds of materials. The students were working on individual projects but were allowed to support each other in the tasks.

The subjects of the interview were two experienced teachers, a male and a female. They did not have any special training in the subject Teknik, but were, as a majority of Swedish teachers of Teknik, trained in Science and Mathematics and had been practising the new subject for at least ten years. The two teachers and the headmaster of the school were informed of the scope of the study and agreed to participate. The interview was performed in Swedish and relevant parts of the material have been translated into English especially for this chapter.

The interview starts, defining the topic

The subjects were informed of the topic of the study, their own, most subjective criteria for the assessment and grading of the work of the students. This is a crucial part of the design of a Repertory Grid Interview. The interviewee must know the elements well and understand the topic. The construct of an individual change in different contexts and the interview was therefore performed in the room where the teachers used to do preparation and assessment work. A RGT interview is often complemented with other forms of data sampling; in this case audio was recorded.

The elements

The teachers had to be knowledgeable of the artefacts and also of the student and his/her design process. We asked the teachers to select 7-8 different projects from their own class. In order to be able to find most of their constructs and criteria they had to make a stratified selection; some good, some bad, some strange and some traditional artefacts. The elements would of course represent the artefact, their designers and also the process of design. The teacher was asked to put labels on every artefact like "The Box", "TV", "Safe" etc. The elements chosen by the male teacher, LG1, were; Green, Telephone, Diary, Plate, Lighthouse, Car and House.

The eliciting of constructs

The software, Repgrid IV, randomly selected three elements; Diary, Lighthouse, and Plate and asked the eliciting question:
– Can you choose two of this triad of elements, which are in some way alike and different from the other one?

This was the crucial point where the informant used his pattern recognition ability on the three selected elements.
– Yes, the Lighthouse and the Diary

The next question was:
– Now I want you to think what you have in mind when you separate the pair from the other one. Just type one or two words for each pole to remind you what you are thinking about when you use this construct?

The answer to this question is a name for the emergent pole, to the left and a implicit or contrasting pole to the right. Although you should be careful in interfering with the process you must ensure that these names are relevant and intelligible, otherwise you may use a "laddering" technique by asking for an explanation e.g.

– The left pole is- They both meet the standards of the task!
– And the right pole is – It is a failure!

Several studies have showed that this eliciting question could be formulated in different ways and that this has an effect on the construct elicited. The advice from many authors is to concentrate on the similarities of a pair and simply ask for the opposite of this. This process is covered in more detail later in this chapter.

The rating of elements

The two poles of the construct were noted in the program and all the elements were rated accordingly, belonging to different extents to one or other of the poles. We had chosen to use nine levels in the ratings, but five or seven are often used. The result when the construct "Failure-Meets the standards" was used to compare the artefacts was:

        Implicit pole: Failure
                9
                8
                7 =Plate
                6
                5
                4
                3= Telephone and House
                2= Lighthouse, Car and Diary
                1= Green
       Emergent pole: Meets the standards

The construct and the ratings of all the elements were noted in the program and produced the first row of the grid. The eliciting process started all over again with three new elements and eliciting of poles of the new construct. The selection of elements was random but the software makes it possible to select them manually. Some software packages may interfere in the process by analysing "on the go" and noting constructs that seem to be too similar. The program will then ask for laddering procedures to try to sort out the difficulties and involves splitting a compound construct in two.

THE REPERTORY GRID TECHNIQUE

The resulting grid

When nine different constructs had been elicited the process was ended. In the case of the first teacher, LG1, the following constructs were elicited:

Emergent pole	Implicit pole
Traditional	Creative
Meets the standards	Failure
Ugly	Beautiful
Good Craftmanship	Novice
No ideas	Highly inventive
Needs support	Self-confident
High grade	Low grade
Persistent	No Endurance
Low Functionality	Good functionality

Display FS (LG1_eng)
"Assessment of Alarms"

Traditional	9	6	6	1	8	7	7	Creative
Meets the standards	1	3	2	7	2	2	3	Failure
Ugly	8	8	5	1	9	6	7	Beautiful
Good craftmanship	2	2	5	9	1	3	4	Novice
No ideas	9	5	7	1	5	5	4	Highly inventive
Needs support	5	9	4	5	9	9	8	Self-confident
High grade	1	2	5	9	2	6	5	Low grade
Persistent	2	3	5	9	3	5	5	No Endurance
Low functionality	9	7	6	1	4	5	6	Good functionality

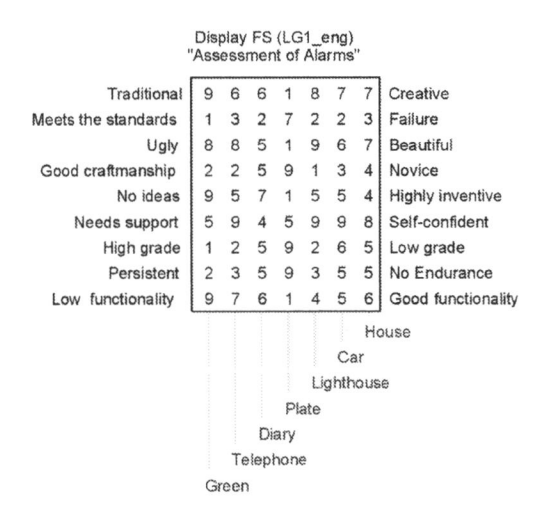

House
Car
Lighthouse
Plate
Diary
Telephone
Green

Figure 1. Display of LG1's constructs in a grid

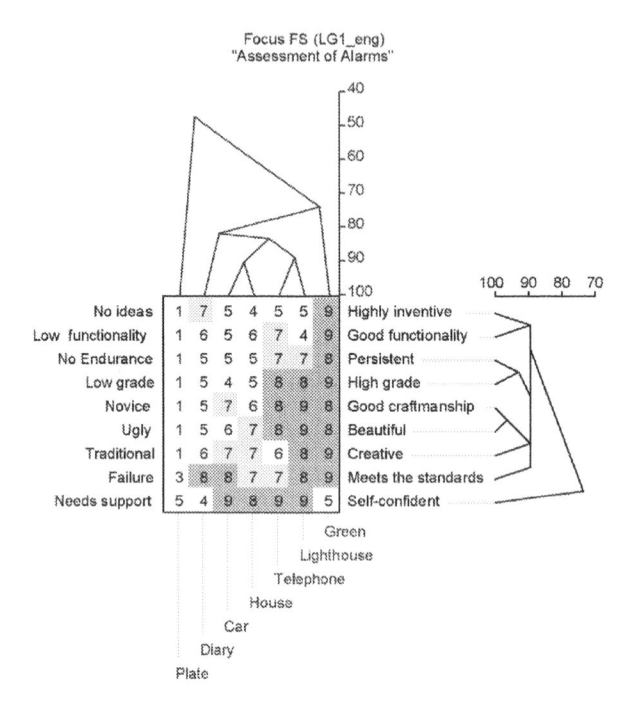

Figure 2. Focus display of LG1

Analysis of the grid

The resulting grid is displayed in Figure 1. The elicited constructs form rows and the elements columns in the grid. To the left is the emergent pole and to the right the implicit pole of a construct. Statistical methods may be used to find similarities or differences in the data but many authors recommend studying the raw data first. Two different tools of analysis in the software package were used. These comprise the Focus display and the Principal Component Analysis Graph.

In the Focus Display, Figure 2, also called hierarchical clustering analysis, the grid has been sorted and rearranged to bring closely matching elements together, and closely matching constructs together. This "focusing" of the structure gives the method its name. The similarity scores of adjacent elements or constructs are provided in numeric form and graphically in a tree structure, a dendrogram. The actual score can be found if you follow the lines from two constructs/elements to the apex and further to the scale.

THE REPERTORY GRID TECHNIQUE

In LG1 we find a strong resemblance of 93%, between the constructs: High grade and Persistent. Even more alike at 96% are the two constructs: Good Craftsmanship and Beautiful. There are strong similarities at 90%, between every construct but: Self-confident-Needs support. This high figure indicates that most of the constructs are relevant to the teacher in his task of assessing. However, you cannot be sure that constructs are not missing in this kind of analysis .

Focusing on the columns, the Elements, we find that the House and the Car are rated as very similar and show a similarity of 91%. The Lighthouse and the Telephone are also rated in a similar way but one of the elements stands out, The Plate. As can be observed, this is obvious from the numbers in the grid and the Focus Display is just a convenient way to show variance graphically to help the researcher get a first view of similarities or differences.

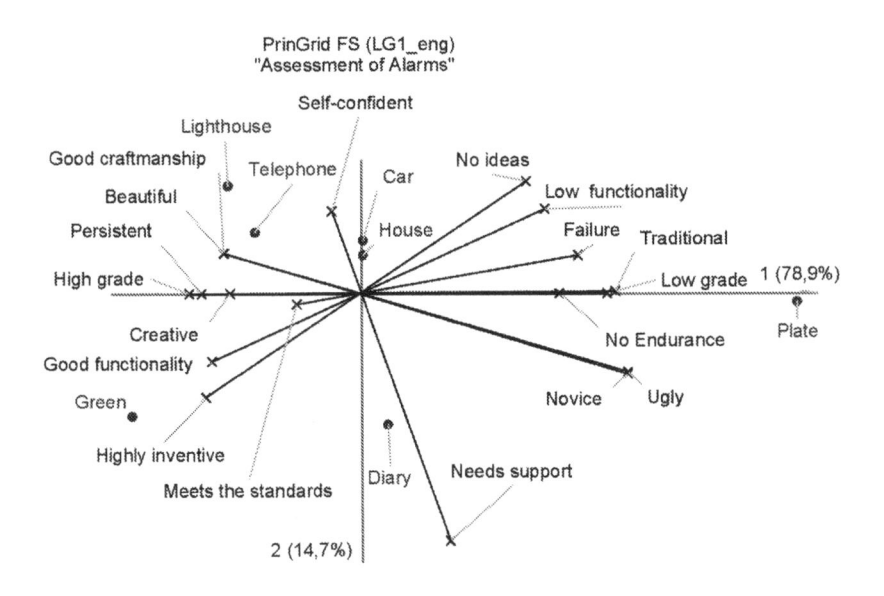

Figure3. Principal Component Analysis of LG1

The Principal Components Analysis identifies distinct patterns of variance on figures in a grid, following the procedure which works out the extent to which the ratings in each row are similar to each other. Iteratively, it attributes as much of the total variance to each distinct pattern, component, as possible. If the total of variance accounted for is more than 80% two components will suffice for the analysis and this is supported by most software packages. The grid has been treated as if the elements were points plotted in an n-dimensional space defined by the constructs as axes centred on the means of the elements. The data has then been rotated to spread the elements out as much as possible in a 2-dimensional plot.

The first principal component identified is plotted as a horizontal dotted line with the percentage of variance printed at the right end. The second component is plotted vertically as a y-axis in a Cartesian system of coordinates. The constructs are plotted as straight lines whose angles with respect to each principal component reflect the extent to which it is represented by the component. The length reflects the amount of variance in the ratings on that construct. The elements can be positioned along each component-axis and the distances between elements will reflect their ratings according to the set of constructs.

In this PCA-graph the first component accounts for 78.9% of the variance and together with the second, 14.7%, will identify 93.6% of the variance in the data. Around the x-axis several important constructs are clustered: High - Low grade; Traditional - Creative; Persistent - No Endurance are more or less the same as the first principal component but: Functionality; Inventiveness; Meeting the standards; Beautifulness; and Craftmanship do form sheaves with small angles around the principal component. They all could be considered important for the grading process. The outstanding construct of Self confident - Needs support has a relatively small angle towards the second component and a perpendicular angle to the other constructs, which would indicate that it is another dimension in the assessment system of this teacher, independent of the others.

The elements are clustered in two parts of the graph. The Plate is on an extreme position completely reflecting its extreme ratings in the grid. Similarly assessed constructs are close to each other in respect to the constructs.

The second teacher in the study, LG2, cooperated in a class with LG1 and by chance selected three identical projects/elements for the grid. You may, as the researcher, select common elements for the subjects of the study but according to the literature you are advised to do this in close cooperation with the interviewee.

THE REPERTORY GRID TECHNIQUE

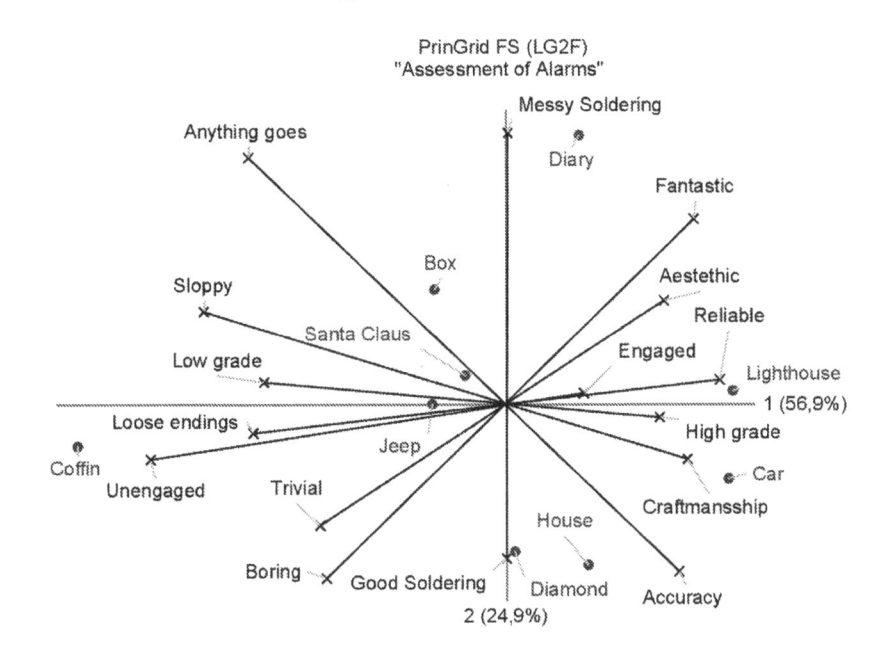

PrinGrid FS (LG2F)
"Assessment of Alarms"

[Figure: PCA plot with constructs including Anything goes, Sloppy, Low grade, Loose endings, Unengaged, Trivial, Boring, Good Soldering, Messy Soldering, Diary, Fantastic, Aesthetic, Reliable, Engaged, Lighthouse, High grade, Car, Craftmansship, Accuracy, Diamond, House, Jeep, Santa Claus, Box, Coffin. Axes: 1 (56,9%) and 2 (24,9%)]

Figure4. Principal Component Analysis of LG2

In LG2's PCA the two principal components accounts for 81.8% of the variance and the constructs are more evenly spread in the graph than is the case with LG1. Strong coherence to the first principal component could be found in the following constructs: High grade - Low grade; Reliable - Loose endings; Engaged - Unengaged; and Craftsmanship – Sloppy. The second component seems to be very similar to the construct. Good -Messy Soldering. If you compare the grids of these two teachers you will find common criteria for assessment but the weighting differs. Both of the teachers are positively grading Aesthetic features but LG1 is giving it higher weight; the constructs Beautiful and Good Craftsmanship are close to the first component and High grades. In LG2 the constructs Fantastic and Aesthetic are more remote. LG2 seems to connect Craftsmanship more to Accuracy. Engagement or Persistency seems to be of importance in both teachers and so is the Functionality/Reliability.

LG1 uses the construct Creative in a way that differs from the use of Inventiveness, which is interesting.

The study this example is drawn from included grid-sessions with teachers from other subjects in the curricula: Essay writing, Web design, History, Philosophy and others. The research question was: Are there common constructs or criteria in teachers' assessment of creative work in different subjects?

Discussion

Kimbell et al explored how one might categorise creative / innovative work (Kimbell, 2004; Kimbell et al., 2006). Teachers, advisers, examiners and other experts in Technology Education were consulted and a list of categorising words was derived, here listed in priority order:

> Exciting, unusual, different, novel, risky, bending the rules, brave, determined, marketable, professional, 'wow', confident, powerful, and unique.

Most of the categories deal with having ideas, growing ideas and proving ideas.

We can identify several of Kimbell's "buzzwords" in the constructs of the two teachers; they are using, on the "positive" pole, abilities as Creative, Highly innovative, Beautiful, Self-confident, Accuracy, Aesthetic, Craftsmanship and Fantastic. These criteria will of course grade students and their products but it is of little value in giving feedback and showing routes to a progression in the development of their creative skills. But there are some constructs; Persistent, Self-confident and Engaged that could be of some value. They deal more with the process and less with the product. (Lindström, 2005) They are dispositions, habits of mind or habits of work that are identified in the students' behaviour.

It could be that teachers in the fine Arts know more about the design and creative process in their field than technology teachers. They may even be artists themselves. The continuing study of teachers assessing written essays seems to indicate just that. Experienced examiners seem to use more qualitative criteria than novices. Probably they have a much larger base of exemplars to relate to. Several of the criteria Lindström identified with teachers in Art and Crafts education seems to be used by teachers in other subjects involved in "making". This will be studied further.

MERITS AND PROBLEMS WITH REPERTORY GRID TECHNIQUES

The Repertory Grid Technique is a type of structured interview which evolved from Kelly's Personal Construct Theory (1955). Its development was Kelly's attempt to present a method of data collection which, first, focused on the individual rather than large groups, as typical of correlation studies of his time. His original technique was the Role Construct Repertory test which was used to investigate role relationships in clinical settings, namely between patients and their families, friends and others, and for assessing relationships between a patient's constructs about people. It needs to be clarified, however, that the Repertory Grid is not a test, but a methodology involving highly flexible techniques and variable application. Thus, although its original use was to investigate constructs about people, subsequent application has included inanimate objects, events, situations and abstract ideas. Many researchers and writers of handbooks in methodology have explained these diverse uses (Cohen & Manion, 1994; Fransella, 2003; Fransella & Bannister, 1977; Jankowicz, 2004).

Topic and examples

People have constructs about anything and everything, however, a RGT is always conducted about a particular topic with the intention to find constructs a subject use in making sense of a particular realm of discourse, that particular slice of their own personal experience. The administration of repertory grids often involves the use of one or more examples as an illustration for participants. Descriptive examples elicit more personally revealing, as opposed to factual, construct dimensions. Recent use of the technique has shown significant differences following from sometimes-subtle variations in repertory grid procedures.(Reeve, Owens, & Neimeyer, 2002)

Elements

Elements may be generated in several ways; you may supply them to the interviewee in the form of a list of people, incidents, actual artefacts, scenes from a video film etc. They could be selected together with the interviewee in a discussion about the topic at hand. You could define a pool of elements and let the person write down: six leisure activities, eight typical parts of a lesson, seven effective teachers etc. Kelly used to provide role titles; "A friend, someone you don't like, your wife, yourself, the ideal self" and asked the interviewee to select specific people he know that would fit the roles. Elements are the important clues that are going to facilitate the elicitation of personal and tacit knowledge from the person in focus. They should be:

- Homogenous, from the same category or dimension

- Representative, covering all aspects of the topic under scrutiny

- Unambiguous, specific, simple and readily understood

- As short as possible, eight to ten elements are quite adequate for most purposes (Easterby-Smith, 1980, 9).

Constructs

Constructs can be generated in several ways, they can be:
− Elicited from triads or dyads.
− Supplied by the researcher, but there is always the danger that the grid becomes inflexible like an attitude questionnaire, with the investigator's world being imposed upon the subject.
− Combine elicited and supplied constructs. Eliciting may take a long time. Supplying a few constructs after eliciting can speed up the process. Certain common constructs may be useful when you want to compare individuals;

examples of supplied construct are Bad-Good, Novice-Expert, Relevant-Irrelevant.
- Non-verbal constructs may be produced simply by sorting element cards into
- different piles according to whatever scheme seem to fit best. These dimensions need not be expressed. The position of each card is noted at the end of each successive sort, and this provides the basis for a grid matrix.
- Laddering is a process of generating additional constructs from existing ones.
- It's done by asking why particular construct poles are important, or by asking for further elaboration of an existing construct.

Normally it is advisable to avoid constructs, which are:
- Concrete (distinctions based on factual attributes).
- Impermeable (can only be applied to a tiny portion of the range of elements).
- Vague (e.g. is OK—not so good).
- Generated by the role title (e.g. is an effective manager—not so effective).

Eight or ten constructs are quite adequate for most purposes (Easterby-Smith, 1980, 9).

Methods of elicitation

The literature identifies systematic differences between Kelly's "difference" and "opposite" methods of personal construct elicitation. The "difference method" has been the standard procedure for construct elicitation and is the single-most commonly used method of construct elicitation. The interviewee is presented with three randomly selected elements (e.g. people), and asked to "identify any two people who are alike in some way, yet different from the third." A subject, when restricted to finding a similarity prior to stating a difference, occasionally is unable to respond. When they do respond there is often a contrast that is not genuinely bipolar. The method has also been said to develop several "bent" (i.e. non antonymous) constructs. In the "opposite" method the interviewee is asked to "identify any two people who are alike in some way." Once that characterization is done the person is then asked to identify the "opposite" of that characterization. This "opposite" then forms the implicit pole, thereby assuring the bipolarity of the construct. The method produces significantly less complex and differentiated personal construct systems. A new method "the Contrast method" yielded personal constructs that were more genuinely bipolar, but without incurring the greater negativity associated with the contrast poles of constructs elicited by way of the opposite method. (Neimeyer, Bowman, & Saferstein, 2005; Yorke, 2001)

The methods above have all been using three elements in the eliciting process but the use of two elements, a dyad, has also been reported in the literature. The task seems to be easier for the interviewee who only has to detect a similarity or a difference in the dyad. On the other hand it has been reported that the constructs produced are not so cognitively complex as when elicited in a triadic fashion. (Caputi & Reddy, 1999)

Reliability and validity

This is a qualitative method and the result in terms of the grid with lots of numbers may lead us to think we have quantitative data. We must remember that the data are the result of a sorting process by the interviewee and show relations between elements, probably not linear and susceptible to the selection of elements and the interview situation. Too much statistical processing of this "soft" data may distort information. The elicited names and labels on the individuals constructs are very subjective and we must take great care to interpret them "correctly".

The theory underlying RepGrid comes from Personal Construct Psychology, where humans are regarded as changing beings. This means that the consistency between systems of constructs, over time, can be low. The subject may change his attitudes towards certain elements. (Gathercole, Bromley, & Ashcroft, 1970)

The validity of the technique in the terms of PCP is its capacity to enable us to elaborate our construing, our ability to anticipate. Ultimately validity refers to the way in which a mode of understanding enables us to take effective action. (Fransella & Bannister, 1977) In this respect Repertory Grid has proven its utility.

Reflective scrutiny

Neimeyer (2002) advocates a position of critical reflection in relation to common adaptations of repertory grid procedures. The fact that even subtle procedural variations can register a substantial impact on the content and structure of a person's Personal/Construct System highlights the responsibility that researchers and practitioners have to understand their own contributions to the grid outcomes that they interpret. (Neimeyer, 2002)

The divorce of Repertory Grid Technique from the theory

The Personal Construct Psychology is firmly based in the area of general cognitive processes, but is also applicable to the individual concepts and their relationship to the solution, as well as an individual's method of progressing towards a solution… and is particularly useful in approaching an understanding of covert behaviour in specific activities, such as designing. (Jerrard, 1998)

Fransella & Bannister note that a lot of studies using RepGrid do not relate to the Personal/Construct Psychology and they suggest that Kelly's theories should be understood in every use of the method. (Fransella & Bannister, 1977) In the Journal of European Industrial Training, important key assumptions with implications for the use of Rep Grid were listed (Easterby-Smith, 1980). These comprise:

> A person's processes are psychologically channelised by the ways in which he anticipates events. (Fundamental Postulate)

> Persons differ from each other in their construction of events. (Individuality Corollary)

A person chooses for himself that alternative in a dichotomised construct through which he anticipates the greater possibility for extension and definition of his system. (Choice corollary)

A person's construction system varies as he successively construes the replication of events. (Experience Corollary)

To the extent that one person construes the construction processes of another, he may play a role in a social process involving the other person. (Sociality Corollary)

Kelly designed the PCP half a century ago and did not have access to the results of modern cognitive science. His work is still used in clinical psychology and could certainly be of value as a theoretic framework for studies in behavioural and attitudinal change. Kelly was not aware of theories of the dual minds, implicit learning and implicit knowledge. He did not have access to results from experimental psychology on familiarity and recognition. In this perspective there are probably other theoretical frameworks that could be utilised to understand the Repertory Grid results.

Merits and advocates

In spite of the shortcomings of the method there are many advocates for the Repertory Grid Technique, a typical example is Mazhmdu:

Many studies in clinical, management and educational settings indicate that the repertory grid is now a well-established diagnostic and research tool. Its idiographic nature encourages the interviewee to use his or her own words when discussing issues of personal importance... It also provides information as to an individual's perceptual field, consequently promoting detailed exploration of personal meaning with a public record compiled easily. The grid provides for the analysis of relationships between constructs and between elements and for analysis of change not only within the same individual but also between individuals over time. Observer bias is reduced almost to zero and objectivity is maximized... The discipline involved in the application of the techniques ensures that each interview is structured and is truly constructive. The interviewers/observers are forced to keep quiet, thereby minimizing their input, while the rigour of the compare and contrasting techniques ensures that the interviewees elaborate at length their understanding of their perceptions. Additionally, the conversational format of the technique offers itself as a tool, which is simple and enjoyable for the interviewees and does not provoke anxiety in them, while being reassured that their own opinions are being sought, so that there are no right or wrong answers. The methodology itself is flexible, elicits both qualitative and quantitative data that are open to a variety of analyses, and its overall potential as an interpretative framework explains why it is generally regarded as a catalyst for change. On the basis of the above explanation, it seems clear that the repertory grid research technique offers a number of fundamental advantages to the researcher. It is also evident that many of the limitations outlined can be overcome and therefore reasonable to suppose that, on balance, the repertory grid is a fruitful technique that should be widely adopted for research purposes throughout nurse education and practice (Mazhmdu, 1992, 605-606).

EXAMPLES OF STUDIES USING THE REPERTORY GRID TECHNIQUE

Assessing creativity

One of the areas where Rep Grid has been used was in the exploration of Design and Creativity. (Jerrard, 1998; Quinn, 1980) Two recent studies show its utility.

In a large study about assessing creative development in Art education Lindström et al (1999) attempted to find criteria or descriptions of general abilities used by teachers to assess creative work. .After a thorough literature survey, several experts who were teachers but also artists and craftsmen, were interviewed with the Repertory Grid and other techniques. Elements in the RGT-interview process were artefacts of fine metal craft. With the help of elicited constructs, laddering and deep interviews of the experts, a list of important factors/criteria for the development of creativity was devised:

Product criteria:

1. Visibility of the intention/ Goal fulfilment

2. Colour, form and composition/ Visual qualities

3. Craftsmanship/ Technical skill

Process criteria:

4. Investigative work/ Persistence in the pursuit

5. Inventiveness/ Imagination and risk-taking

6. Ability to use models/ being able to learn from others

7. Capacity for self-assessment/ knowing one's strengths/weaknesses

Others:

8. Overall judgement

Each criterion was described in four steps, notifying a developmental, progressional change from the behaviour of a novice to that of an expert. The levels were described in a narrative way as a rubric, to make it possible for a teacher to recognise the behaviour of a particular student.

The following rubrics were used for scoring Craftsmanship:

1. The pictures show little or no ability to use materials and techniques.

2. The pictures suggest a certain ability to use materials and techniques, but there are serious deficiencies in the execution.

3. The pictures show an ability to use materials and techniques to achieve the desired visual effects, but this is applied in a rather stereotyped way.

4. The pictures show a good and flexible mastery of materials and techniques and are consistently of high technical quality.

The following rubrics were used for scoring Capacity for self-assessment:

1. The student cannot point out the strengths and weaknesses of her own work or distinguish between works that are successful and those that are less successful. She has no opinions about her fellow students' pictures.

2. With some assistance, the student can point out the strengths and weaknesses of her own work and distinguish between works that are successful and those that are less successful. Opinions about her fellow students' pictures are confined to simple value judgements (good/bad, like/don't like).

3. The student is generally able to see merits and shortcomings in her work and can select sketches, drafts, and works which illuminate her own development. She can pass varied judgements on her fellow students' pictures.

4. The student can clearly see merits and shortcomings in her work and can select sketches, drafts, and works which illuminate her own development. She can also give reasons for her judgements and explain why things turned out as they did. She can pass varied judgements on her fellow students' pictures and is able to give constructive criticism.

There is an obvious difference between the criteria of excellence that were generated by this and following studies and those checklists that are often found in textbooks. The latter lists components that should be present in a product or performance, while the interviewees in this study tried to define a set of more general dispositions or key competencies. The typical textbook items are, at best, indicators of such "habits of mind". The interviewees' process criteria, in particular, add up to a culture of learning rather than a list of specific skills (Lindström, 2001).

Lindström found high agreement between class teachers and co-assessors in ratings of both the students' visual results (product criteria) and their approach to work (process criteria). In almost 3,100 comparisons between class teachers and the co-assessors from another school, there was 78 per cent agreement (= 2 steps on a twelve-grade scale). Given that other discrepancies between the two assessors were small and indicate an approximately normal distribution, this may be regarded as a satisfactory result (Lindström et al., 1999).

Another study by Hassenzahl & Wessler, (2000) attempted to capture Design Space From a User Perspective:

The design of an artefact (e.g. software system, household appliance) requires a multitude of decisions. In the course of narrowing down the design process, "good ideas" have to be divided from "bad ideas". To accomplish this, user perceptions and evaluations are of great value. The individual way people perceive and evaluate a set of prototypes designed in parallel may shed light on their general needs and concerns. The authors assume that the personal constructs (and the underlying topics) generated as a reaction to a set of artefacts mark the artefacts' design space from a user's perspective and that this information may be helpful in separating valuable ideas from the not so valuable... In general, the Repertory Grid Technique proved to be a valuable tool in exploring a set of artefact's design space from a user's perspective (Hassenzahl & Wessler, 2000, 441).

Educational research

Course evaluations, learning outcome, teachers practice and conceptual change are areas where RepGrid also has been used extensively. (Hoogveld, Paas, Jochems, & Van Merrienboer, 2002; Karppinen, 2000; Pill, 2005; Smith, Hartley, & Steward, 1978) A typical example of this one; a study probing students' unique view of energy; the need for revision of the concept of an average student; information on the extent to which school science ideas about energy had been translated into students' everyday working knowledge and constructs elicited as a basis for interviews (Fetherstonhaugh, 1994).

Tacit knowledge, business and management

Ever since the Journal of European Industrial Training devoted a full issue to Repertory Grid (Easterby-Smith, 1980) there have been a multitude of studies and uses of the technique in management and business. (Chao, Salvendy, & Lightner, 1999; Crowther, Hartnett, & Williams, ; De Leon & Guild, 2003; Easterby-Smith, Thorpe, & Holman, 1996)

Attitudes

Kelly's own area of research was in psychology with a special interest in personal attitudes and emotions. His method is still used to explore people's feelings, prejudices and preferences (Honey, 2001; Parkinson & Lea, 1991).

A good example of this is an article, "Understanding public attitudes to technology", (Frewer, Howard, & Shepherd, 1998) where the authors states in the abstract;

The social context, which surrounds technology, is likely to be one of the most important determinants of its future development and application. The application of repertory grid techniques in conjunction with generalized Procrustes analysis identified important psychological constructs which determine attitude. A larger survey study examined the reliability and predictive capacity of these items in quantifying attitudes to technology. Factor analysis identified two subscales, which appeared to assess perceptions of technological risk and benefit... An inverse relationship between perceived risk and benefit was found, consistent with previous research in risk perception... A major problem in using researcher generated characteristics in scale development is that it is possible that some key determinants of public attitudes are not recognized by the researcher, and so no attempt is made to incorporate them into questionnaire design. Against this, the researcher may decide that particular elements are important in attitude formation when they are not. Even if items are not highly salient to people, they will still produce ratings which will then be incorporated into the subsequent model. What is required is a method where respondents generate their own descriptions of concerns associated with a particular 'target' (for example, a technology), which will avoid some of the problems linked to experimenter generated characteristics... The use of the repertory grid method permits responses to be focused within the hazard domain without imposing external experimenter determined risk characteristics on data generation (Frewer et al., 1998, 222).

CONCLUDING WORDS

Repertory Grid Technique is an interview technique that utilises an individual's ability to compare elements to elicit attitudes, category making, assessing criteria and probably some personal tacit knowledge. It is a qualitative method in which statistical methods may be used to enhance analysis. As in most qualitative methods, questions of relevance, reliability and validity must be addressed, usually with complementary interview and data sampling methods. The method is sensitive to the design of the interview but so are many alternative methods e.g. the bias that the researchers often impose on an interviewee in a traditional interview is reduced in RGT. The RGT is a fruitful technique that could be widely adopted for research and development purposes throughout design & technology education and practice.

REFERENCES/BIBLIOGRAPHY

Caputi, P., & Reddy, P. (1999). A comparison of triadic and dyadic methods of personal construct elicitation. *Journal of Constructivist Psychology, 12*, 253-264.

Chao, C.-J., Salvendy, G., & Lightner, N. J. (1999). Development of a methodology for optimizing elicited knowledge. *Behaviour & Information Technology, 18*(6), 413-430.

Cohen, L., & Manion, L. (1994). *Research methods in education* (4 ed.). London: Routledge.

Cronin, M. A. (2004). A Model of Knowledge activation and insight in problem solving. *Complexity, 9*(5), 17-24.

Crowther, P., Hartnett, J., & Williams, R. (ND).. *Teaching repertory grid concepts for knowledge acquisition in expert systems: An interactive approach*. Unpublished manuscript, Department of Applied Computing and Mathematics. University of Tasmania.
De Leon, E. D., & Guild, P. D. (2003). Using repertory grid to identify intangibles in business plans. *Venture Capital, 5*(2), 135-160.
Dreyfus, H. L., & Dreyfus, S. E. (1986). *Mind over machine*. Oxford: Basil Blackwell Ltd.
Easterby-Smith, M. (1980). How to use repertory grids in HRD. *Journal of European Industrial Training, 4*(2), 2-32.
Easterby-Smith, M., Thorpe, R., & Holman, D. (1996). Using repertory grids in management. *Journal of European Industrial Training, 20*(3), 3-30.
Education, D. O. (2000, 20030515). The Swedish curriculum for technology education, from http://www3.skolverket.se/ki/eng/comp.pdf
Epstein, S., Lipson, A., Holstein, C., & Huh, E. (1992). Irrational reactions to negative outcomes: Evidence for two conceptual systems. *Journal of Personality and Social Psychology, 62*(2), 328-339.
Ericsson, A. K., & Charness, N. (1997). Cognitive and developmental factors in expert performance. In P. J. Feltovich, K. M. Ford & R. R. Hoffman (Eds.), *Expertise in context* (pp. 3-41). Menlo Park: AAAI Press / The MIT Press.
Fetherstonhaugh, T. (1994). Using the repertorygrid to probe students' ideas about energy. *Research in Science & Technological Education, 12*(2), 117-128.
Fransella, F. (2003). *International handbook of personal construct psychology*. Chichester, West Sussex, England: J. Wiley & Sons.
Fransella, F., & Bannister, D. (1977). *A manual for repertory grid technique*. London: Academic P.
Frewer, L. J., Howard, C., & Shepherd, R. (1998). Understanding public attitudes to technology. *Journal of Risk Research, 1*(3), 221-235.
Gaines, B. R., & Shaw, M. L. G. (2003). Personal construct psychology and the cognitive revolution, from http://pages.cpsc.ucalgary.ca/~gaines/reports/PSYCH/SIM/index.html
Gathercole, C. E., Bromley, E., & Ashcroft, J. B. (1970). The reliability of Repertory Grids. *Journal of Clinical Psychology, 26*(4), 513-516.
Hassenzahl, M., & Wessler, R. (2000). Capturing design space from a user perspective: The repertory grid technique Revisited. *International Journal of Human-Computer Interaction, 12*(3&4), 441-459.
Hillier, Y. (1998). Informal practitioner theory: Eliciting the implicit. *Studies in the Education of Adults, 30*(1), 35-52.
Honey, D. (2001). *The repertory grid in action: How to use it to conduct an attitude survey*.
Hoogveld, A. W. M., Paas, F., Jochems, W. M. G., & Van Merrienboer, J. J. G. (2002). Exploring teachers' instructional design practices from a systems design perspective. *Instructional Science, 30*, 291-305.
Jankowicz, D. (2004). *The easy guide to repertory grids*. Chichester, West Sussex, England; Hoboken, N.J.: Wiley.
Jankowicz, D., & Hisrich, R. D. (1987). Intuition in small business lending decisions. *Journal of Small Business Management*, July, 45-53.
Jerrard, R. (1998). Quantifying the unquantifiable: An inquiry into the Design Process. *Design Issues, 14*(1), 42-53.
Karppinen, S. (2000). Repertory grid technique in early childhood as a tool for reflective conversations in arts education. Paper presented at the *10th European Conference on Quality in Early Childhood Education*, London.
Kelly, G. A. (1955). *The psychology of personal constructs*. New York: Routledge.
Kimbell, R. (2004). Assessment in design and technology education for the Department of Education & Skills, UK. Paper presented at the *3rd Biennial International Conference on Technology Education Research*, Brisbane.

Kimbell, R., Bain, J., Miller, S., Stables, K., Wheeler, T., & Wright, R. (2006). *Assessing design innovation*. London: Goldsmiths College, University of London.

Lieberman, M. D. (2000). Intuition: A social cognitive neuroscience approach. *Psychological Bulletin, 126*(1), 109-137.

Lindström, L. (2001). *Criteria for assessing students' creative skills*. Unpublished manuscript, Rotterdam 26-29 September.

Lindström, L. (2005). *Assessing craft and design: Conceptions of expertise in education and work*.Unpublished manuscript, Stockholm.

Lindström, L., Ulriksson, L., & Elsner, C. (1999). *Portföljvärdering av elevers skapande i bild.(Portfolio Assessment of Students' Creative Skills in the Visual Arts) (No. 99:488)*. Stockholm: Skolverket / Liber Distribution.

Mawson, B. (2003). Beyond 'the design process'; An alternative pedagogy for technology education. *International Journal of Technology and Design Education,13,* 117-128.

Mazhmdu, G. N. (1992). Using repertory grid research methodology in nurse education and practice:a critique. *Journal of Advanced Nursing, 17*, 604-608.

Middleton, H. (2002). Complex problem solving in a workplace setting. *International Journal of Educational Research, 37*, 67 –84.

Middleton, H. (2005). Creative thinking, values and design and technology education. *International Journal of Technology and Design Education,15*, 61-71.

Neimeyer, G. J. (2002). Towards reflexive scrutiny in repertory grid methodology. *Journal of Constructivist Psychology, 15*, 89-94.

Neimeyer, G. J., Bowman, J. Z., & Saferstein, J. (2005). The effects of elicitation techniques on repertory grid outcomes: Difference, opposite and contrast methods. *Journal of Constructivist Psychology, 18*, 237–252.

Nightingale, P. (1998). A cognitive model of innovation. *Research Policy, 27*, 689–709.

Parkinson, B., & Lea, M. (1991). Investigating personal constructs of emotions. *British Journal of Psychology, 82*, 73-86.

Petroski, H. (1996). *Invention by design*. Cambridge: Harvard University Press.

Pill, A. (2005). Models of professional development in the education and practice of new teachers in higher education. *Teaching in Higher Education, 10*(2), 175-188.

Polanyi, M. (1966). *The tacit dimension*. Gloucester,Mass.: Doubleday & Company, Inc.

Quinn, E. (1980). Creativity and cognitive complexity. *Social Behavior and Personality, 8*(2), 213-215.

Reber, A. S. (1989). Implicit learning and tacit knowledge. *Journal of Experimental Psychology: General, 118*(3), 219-235.

Reeve, J., Owens, R. G., & Neimeyer, G. J. (2002). Using examples in repertory grids: The influence of construct elicitation. *Journal of Constructivist Psychology, 15*, 121-126.

Schon, D. A. (1987). *The reflective practitioner*. London: Temple Smith.

Sloman, S. A. (1996). The empirical case for two systems of reasoning. *Psychological Bulletin, 119*(1), 3-22.

Smith, M., Hartley, J., & Steward, B. (1978). A case study of repertory grids used in vocational guidance. *Journal of Occupational Psychology, 51*.

Stevenson, J. (2003). Expertise for the workplace. In J. Stevenson (Ed.), *Developing vocational expertise* (pp. 3-25). Crows Nest Australia: Allen & Unwin.

Sun, R., Slusarz, P., & Terry, C. (2005). The interaction of the explicit and the implicit in skill learning:A dual-process approach. *Psychological Review, 112*(1), 159-192.

Williams, P. J. (2000). Design: The only methodology of technology. *Journal of Technology Education, 11*(2), 48-60.

Yorke, M. (2001). Bipolarity ... or not? Some conceptual problems relating to bipolar rating scales. *British Educational Research Journal*, *27*(2), 171-185.

Lars Björklund
Swedish National Graduate School in Science and Technology Education Research
University of Linköping
Sweden

IVAN CHESTER

RESEARCHING EXPERTISE DEVELOPMENT IN COMPLEX COMPUTER APPLICATIONS

INTRODUCTION

This chapter describes research in the three-dimensional computer aided design (3D-CAD) context outlining the factors that need consideration when choosing appropriate knowledge elicitation techniques and research methodologies for the types of learner/computer interactions that occur. In order to undertake this research new knowledge elicitation techniques need to be employed that will help to gather the detailed data that enables generation of an understanding of the cognitive processes and human-software interactions involved.

The teaching of Technology Education progressively involves greater use of computer technology. Computers are now commonly used to enable students to control devices such as lathes, mills and robots, to acquire data for input into the process of improving the solutions to problems, to design new products through the use of 3D-CAD, to design and test electronic, hydraulic and pneumatic circuits, to store and present information, to access information via the internet, and as a means by which learning material is presented, tested and recorded through computer based instruction. However, despite the plethora of new software with which both technology students and teachers interact on a daily basis little research has been undertaken that seeks to understand, in any systematic way, the nature of the cognitive interaction involved, or the manner in which students can progress from using the software at a relatively simple or superficial level to efficient or expert software use.

The research outlined in this chapter formed part of a larger study designed to devise and test methods that may improve the use of expert strategies by novice users of 3D-CAD. This task involved identifying expert 3D-CAD users and then identifying the cognitive strategies used by those experts. The research therefore needed to be grounded in the cognitive psychology literature on both expertise and knowledge elicitation. Expertise, in order that the data collected reflected expert use of the software and was therefore valid, and knowledge elicitation so that the data collected was sufficiently rich and reliable for effective analysis of the cognitive skills involved. This approach is supported by Taylor (in Evans, 1991 pp. 170-171) who maintains that the first stages of any research into the novice-expert shift should entail;

> Identification of the cognitive skill performance that is the ultimate objective of instruction;

Analysis of expert performance of this skill in terms of the content, structure and organisation of the underlying declarative knowledge base;

Design of instrumentation to measure salient aspects of this declarative knowledge base and associated actual cognitive skill performance. (Taylor, in Evans, 1991, pp. 170-171).

The first stage of this process involves the identification of individuals for inclusion in the research in order to ensure valid data is collected. In the case of the type of research being discussed in this chapter, research into the expert use of computer software, it becomes even more critical as the knowledge being sought relies on the initial subject(s) being not just users but expert users.

Expertise

Matlin (2005), Mieg (2001) Sternberg (1990) and Sternberg and Grigorenko (2003) argue that experts, unlike novices or advanced novices, possess high levels of domain knowledge, well developed cognitive structures in the form of schemas, the ability to recognize large numbers of domain specific patterns, a propensity to represent problems at a deeper more structured level and prominent use of executive control or metacognition during problem solving. It is hypothesised that expertise in 3D-CAD exhibits similar characteristics and it is therefore critical that experts be identified, as other users may not possess the cognitive structures that characterize expertise in the 3D-CAD domain.

There are potential difficulties associated with the identification of computer software expertise, and 3D-CAD expertise in particular. Mieg (2001) supported by Charness and Schultetus (1999) define expertise as "consistently superior performance on a specified set of representative tasks for the domain that can be administered to any subject" (Mieg, 2001, p.76). The criteria suggested for expert selection include distributions on tests (Charness & Schultetus, 1999), peer nomination (Mieg, 2001), or experience (Charness & Schultetus, 1999). For 3D-CAD, as with many other areas of complex computer software use, these methods present difficulties due to the absence of tests across the various software applications and the lack of data available for peer nomination through related professional associations. Charness and Schultetus (1999) also point out that experience does not always equate with expertise, a position supported in the CAD research of Bhavnini and John (1997) who found that experienced CAD users often persisted in the use of sub-optimal strategies. Hoffman, Shadbolt, Burton and Klein (1995) and Schraagen, Chipman and Shalin (1999) observe that there is a scarcity of true experts, but make the observation that "articulate experts with recent experience in both performing and teaching the skill are particularly useful" (Schraagen, Chipman & Shalin, 1999, p.6).

The criteria that may be effective in the identification of experts therefore are: superior performance on tests, peer nomination, experience and recent teaching experience (Charness & Schultetus, 1999; Mieg, 2001). However, test performance has been found to be impractical in the 3D-CAD domain. On this

basis, and taking cognizance of the expertise literature, criteria for the identification of potential research subjects for this 3D-CAD research were developed as follows:
- They have a number of years experience in the domain.
- They currently use 3D-CAD on a regular basis.
- Others regard them as possessing domain expertise.
- They have experience in teaching/training others.

Subject selection

Hoffman, Shadbolt, Burton and Klein (1995) and Schraagen, Chipman and Shalin (1999) observe that there is a scarcity of true experts. The sample size used in this study therefore reflects the difficulty in identifying true three-dimensional solid modelling CAD experts. Using the criteria established above the selection of subjects was undertaken. Two different methods were used in this process. Firstly, subject identification was undertaken via an advertisement in a specialized CAD magazine with nation-wide distribution. The magazine editor was supportive of the research and provided free space for the research description and a call for volunteer participants who would meet the selection criteria. Four subjects were identified through this process, however, data collected from three of these subjects proved unusable as they were unable to complete the modelling process, indicating that they were not as expert as hoped. A second identification process utilised individual CAD vendors known to the researcher. These were contacted in an effort to identify further subjects who would meet the selection criteria. Three additional subjects were identified through this process. The four subjects for the study were therefore considered to meet the selection criteria for expertise in CAD. In addition, these subjects used a range of 3D-CAD programs thus providing data that would be independent of specific software. The processes of eliciting expert three-dimensional solid modelling CAD knowledge from these subjects in order to undertake detailed analysis of the nature of three-dimensional solid modelling CAD expertise are discussed below.

Knowledge elicitation

The processes and validity of data collection, or knowledge elicitation, are critical in the type of research under discussion. Cooke (2000), for example, points out that:

> unlike performance-critical applications such as expert systems, applications like training that go beyond knowledge use to the transfer of knowledge, require more attention to the validity of elicited knowledge (Cooke, 2000, p. 481).

Cooke (2000) provides an overview of knowledge elicitation methods outlining four alternative methods; verbal reports, observations, interviews, and process tracing, noting that each method has inherent advantages and shortcomings. Each of these processes is discussed in the following section.

Verbal reports

Verbal reports involve the subject documenting their cognitive processes either orally or in written form. Ericsson and Simon (1993) propose verbal reports as a suitable method "to study superior performance on specific tasks under controlled conditions and to assess the cognitive processes, knowledge and acquired mechanisms that mediate the superior performance of experts" (p. XXXVIII). Verbal reports may take a number of forms: concurrent oral reporting during the process of performance, often referred to as think-aloud protocols; retrospective reports in the form of either recorded verbal or written reports following performance; or a structured or unstructured interview whereby the researcher seeks to gain insight into cognitive processes via specific questioning of performance (Cooke, 2000). Each of these types of verbal reports is now discussed.

Think-aloud protocols

In the think-aloud process the subject undertakes a specific task (in the case of this research the production of a 3D-CAD model) while at the same time talking out loud saying everything that they would otherwise say to themselves. Think-aloud verbal protocols thus enable the researcher to gain information from the expert while they are in the process of performing the task. While this may be advantageous, there is some concern (Cooke, 2000, Gordon, 1992; Hoffman, Shadbolt, Burton & Klein, 1995) that the process of verbalising may affect task performance and that the information gathered may not provide the necessary insight as many of the expert's cognitive processes may be automated. It is suggested (Cooke, 2000) that concurrent verbalizing can only report current consciousness and not explanations or interpretations. To attempt to do anything else may interfere or in fact change the thought processes of the individual. Thus automated processes or information regarding perception or retrieval may not be collected accurately in this manner but would need to be inferred from the information collected. Rowe (1985) points out "the analysis of 'thinking aloud' protocols does not provide a complete description of the problem solving process. It is, however, a method which permits the externalization of certain covert processes, and thereby provides the investigator with an initial tool" (p. 106).

Ericsson and Simon (1993) outline three levels of verbalisation. Type 1 involves "vocalization of covert articulatory and oral encodings" (p.79); Type 2 "involves description, or rather explication of thought content" (p.79); and Type 3 "requires the subject to explain his thought processes or thoughts" (Ericsson & Simon, 1993, p. 79). These authors provide an analysis of the efficacy of verbal reports as data claiming that verbalizing of Type 1 and Type 2 thoughts is found to have little effect on task solving performance. Where some effects may be evident, they pertain to speed and accuracy, neither of which are considerations in this research. However, Type 3 verbalization, due to the fact that the reasoning processes are articulated, has been found to affect the problem solving process. Ericsson and Simon (1993) therefore claim that it is imperative that very specific instructions are given to participants in order to ensure that only Type 1 and Type 2 verbalizing is undertaken by the subject. The suggested wording of instructions includes;

> I will ask you to talk aloud as you work on the problems. What I mean by talk aloud is that I want you to say out loud everything that you say to yourself silently. Just act as if you are alone in the room speaking to yourself (Ericsson & Simon, 1993, p. 376).

The nature of the instructions given to participants was therefore important in order to ensure that the problem solving process remained valid. The issue of overall validity of data gathered via verbal protocols is discussed below.

Validity of verbal protocols

Ericsson and Crutcher (1991) point out, in relation to the validity of data collected through think-aloud protocols, that it is:

> difficult to seriously argue that the verbal reports are not a valid model for generating the correct answers to the problems. Such a critical claim must assume that the subject used one model to generate the verbal report and a different model to generate the problems solution, when, in fact, the model used to generate the report is sufficient to generate the problem solution as well (Ericsson & Crutcher, 1991, p. 67).

Based upon their extensive review of the literature on verbal protocols as data, Ericsson and Simon (1993) propose six assumptions upon which verbal protocols are based. These are:

> The verbalized cognitions can be described as states that correspond to the contents of STM (short term memory) (p. 221)

> The information vocalized is a verbal encoding of the information in short-term memory (p. 221)

> The verbalization processes are initiated as thought is heeded (p. 222)

> The verbalization is a direct encoding of the heeded thought and reflects its structure (p. 222)

> Units of articulation will correspond to integrated cognitive structures (p. 225)

Pauses and hesitations will be good predictors of shifts in processing of cognitive structures (Ericsson and Simon, 1993, p. 225).

Ericsson and Simon (1993) conclude "the information that is heeded during performance of a task, is the information that is reportable; and the information that is reported is information that is heeded" (p. 167) thus supporting the validity of the collected data. Ericsson and Simon (1993, pp. 171-172) provide a number of means by which the validity of the data may be confirmed when undertaking analysis of the data. The relevant questions are:
- Were the processes in the verbal protocol necessary for the solution of the problem – if so the data is reflecting the cognitive processes involved and not being generated as an independent process? The data could therefore be considered a valid representation of the cognitive processes involved.
- Are the verbalizations relevant to the task?
- Are the verbalizations relating to the preceding ones?
- Are the verbalizations consistent with information previously mentioned – if so they are accessing the same memory?
- Are the verbalizations consistent with perceptual information – not using cues such as eye movement as in prior research but recoding screen capture of image and mouse movement?

If these conditions are met, Ericsson and Simon (1993) argue that validity of the verbal protocol data is established.

Retrospective reports

A retrospective report is one where "the subject focuses on the task until its completion and then recalls, as accurately as possible, the sequence of thoughts that occurred while doing the task" (Ericsson & Crutcher, 1991, p. XVI). In relation to the process of knowledge elicitation of expert 3D-CAD processes there are a number of potential problems with collecting retrospective reports. Ericsson and Simon (1993) point out that for tasks longer than 0.5-10 seconds' duration the sequence of thought will lack completeness and accuracy. As the completion of a 3D-CAD model is often quite a long process, well in excess of the duration suggested by Ericsson and Simon (1993), this presents a potential problem for the validity of the collected data. Cooke (2000) and Rowe (1985) also highlight a number of potential difficulties with retrospective reports that may also affect the validity of data collection in the current research. These are: there may be a difference between the described method and the actual method; the process of recording the description orally may be unfamiliar or rushed resulting in essential information being omitted; and the process of recording the description in written form may be extremely time consuming, imposing a further impost on the experts that they may be unwilling to undertake or may again rush, resulting in incomplete data. Retrospective reports were thus considered problematic as a valid knowledge elicitation method for 3D-CAD.

Interviews

The use of interviews as a method of knowledge elicitation is, according to Cooke (2000) a "direct way to find out what someone knows" (p. 487). Cooke (2000), Gordon (1992) and Hoffman, Shadbolt, Burton and Klein (1995) outline two methods by which interviews may be undertaken, unstructured interviews and structured interviews. Gordon (1992) argues that these methods, in the main, elicit declarative and not procedural knowledge, a position supported in part by Cooke (2000) who explains that considerable skill and training may be necessary in order for interviewers to use unstructured interviews or the highly specific interview methods necessary in order to gain insight into an expert's procedural or strategic knowledge.

Hoffman et al. (1995) describe an unstructured interview as "an open dialogue in which the interviewer asks open-ended questions about an expert's knowledge and reasoning" (p. 134). This technique, according to Cooke (2000) and Hoffman et al. (1995), may enable the researcher to elicit a wide variety of data regarding declarative, procedural and metacognitive knowledge through the inclusion of open-ended (what, why, how) questions. However, the disadvantages of the unstructured technique include the need for interviewer training, the possibility of the interview getting 'off the track', the possibility that the expert may assume that the interviewer has a knowledge of the domain and the exhaustive nature of the data collected that will need transcription and analysis (Hoffman et al., 1995).

Structured interviews, on the other hand, are designed and planned in advance in order to provide a greater focus on domain specific information. They rely on the interviewer having knowledge of the domain but, according to Hoffman et al. (1995), have the advantages that the interview time is reduced and specific declarative and procedural domain knowledge may be elicited. Cooke (2000) and Hoffman et al. (1995) describe a number of forms of structured interview. In a scenario simulation interview, the expert is presented with a problem and the interviewer walks the expert through the problem solving process. The use of questions such as "Why would you do that?", "How would you do that?", "What alternative strategies could you use?" will, according to Hoffman et al. (1995), help to elicit domain procedures and reasoning rules. An alternative structured interview strategy is to use event recall whereby the expert is asked to explain a particular past event or events. These events may be either typical or particularly salient or difficult cases. Hoffman et al. (1995) argue "probed recall of past tough or salient cases can be very effective in revealing experts' knowledge, especially their tacit knowledge and reasoning strategies" (p. 136).

Cooke (2000) posits a number of advantages for the use of verbal descriptive reports for knowledge elicitation. In relation to the elicitation of expert 3D-CAD knowledge the interview method has two advantages. It saves time on the part of experienced users and therefore potentially gains more cooperation, and it provides the opportunity to request experienced users to use an alternative strategy from their initial preference. Analysis of these data enables the researcher to discover whether the strategy chosen initially is linked to the functions available within the software package being used or is independent of them.

However, the use of interviews may have a number of disadvantages. Firstly, there may be a difference between the described method and the actual method of software used that could affect the validity of the study. Secondly, the method of operation described may not actually work within the software environment therefore also affecting the validity of the study. These validity issues may only be reduced if the interviewer has well developed domain knowledge of all the specific software being used. Finally, the description provided may concentrate too much on the surface processes without providing sufficient insight into the automatic processes employed. Schraagen, et al (1999) support this notion in that experts may often not have "direct conscious access to their relevant knowledge and skills" (p. 8).

A number of alternative knowledge elicitation methods utilising verbal reports has been described and their merits and shortcomings in relation to expert three-dimensional solid modelling CAD knowledge discussed. These methods include; think-aloud protocols, written and verbal retrospective reports, and, structured and unstructured interviews. Cooke (2000) and Ericsson and Simon (1993) argue that, provided specific conditions are met, each of these techniques is capable of providing valid data. For this research it was decided to use think-aloud protocols. Think-aloud protocols enable the capture of data during the process of completing a 3D-CAD task avoiding the potential problems associated with retrospective reports.

An additional method, process tracing, may also be applicable in the three-dimensional solid modelling CAD environment. This process is discussed in the following section.

Process tracing

Schraagen et al. (1999) describe the 'process tracing' method as an observational method of data elicitation. Further Cooke (2000) defines process tracing as "the collection of sequential behavioural events and the analysis of the resulting event protocols so that inferences can be made about the underlying cognitive processes" (p. 490). The strength of process tracing, according to Cooke (2000), is that the data are collected, providing the researcher is unobtrusive, in a natural setting, thus enabling the actual expert behaviour to be viewed. Data collection may be undertaken via direct observation and recording by the researcher or more commonly through an alternative method such as video recording. This data collection method may have the advantages, in the case of 3D-CAD expertise, that:
− the actual process of software use is recorded thereby providing a more authentic understanding of the procedures employed,
− the relationship between the software strategy being employed and the particular stage of the process would also be available to the researcher.

It may have the disadvantages that:
- subjects may not have the available technology necessary for the recording process,
- the time and inconvenience involved in establishing the recording process may mitigate against subjects volunteering to be involved in the study,
- subjects may feel constrained due to the recording process,
- the process can be costly,
- rich data associated with lengthy data collection sessions mean that data analysis may be unwieldy, time consuming and interpretation difficult (Cooke, 2000; Schumacher & Czerwinski, in Hoffman, Ed. 1992).

In order that process tracing may be employed as a knowledge elicitation method for this study the disadvantages outlined above needed to be addressed. In addition, the recommendation of Schraagen, et al (1999) to include the "collection of verbal think-aloud protocols while the SMEs [subject matter experts] perform a representative set of task problems" (pp. 8-9) needed consideration. Adopting both methods would combine the benefits of the think-aloud process outlined above with process tracing. This combination would provide a very rich data source and facilitate the linking of verbal and visual data thus enabling each data source to validate the other. The validity of the verbal protocols would be able to be appraised in accordance with the criteria established by Ericsson and Simon (1993), noted earlier: applicability to the solution of the task, relevance to the task and consistency with perceptual information.

In their summary of data collection methods Schraagen, et al (1999) point out that in the choice of a suitable knowledge elicitation technique it is important that the process needs to be "easy and natural for the subjects" (p. 469). In choosing a process tracing method this would require a recording method that took into account the needs of the subjects and the potential logistic difficulties presented to the researcher by having subjects in widely disparate locations. For this and economic reasons video recording was rejected. Following the recommendation of Cooke (1999, p. 501) that "unlike traditional knowledge elicitation and task analytic methods, methods that focus on computer-recorded events can amass data in the background, posing little threat of interference to task performance" a number of computer software options was identified that allowed for the process of recording to be achieved. These included the Lotus product "ScreenCam", Microsoft "Camcorder" and the TechSmith product "Camtasia" that are commonly used for the production of computer-based instruction or product promotion material. In each case the software enables the capture of all computer operations as they occur in real time on the screen and records them in a video format. They all have the additional feature of being able to record sound at the same time in the form of voice-over.

These products were evaluated therefore as having the potential for process tracing. For each, the software could be readily installed on the subject's computer, and a microphone supplied, so that the subject could record think-aloud protocols as they undertake the production of a 3D-CAD model. Once the recording process is commenced the software recedes into the background and remains unobtrusive while the 3D-CAD task is undertaken. All computer actions and think-aloud protocols are recorded without further subject intervention. Once the 3D-CAD task is complete the subsequent video is then saved to a CD-ROM and returned to the researcher for analysis.

Tests of each of the software packages were undertaken to ascertain their effectiveness. While each was capable of performing the necessary function "Camtasia" was chosen as the preferred option for a number of reasons. Firstly, it was the only package useable over a range of computer operating systems. This was an important factor, as subjects working in industry, in particular, often run their CAD software on alternative systems, such as Windows NT, in order to obtain network stability. Secondly, it was also available in a time restricted trial version via the Web, thus overcoming the potential problems associated with software licensing. The trial version could be downloaded to CD and supplied to subjects with instructions for installation. Thirdly, the resultant video file was in a form readable on almost any computer thus enabling ease of analysis. Process tracing using "Camtasia" software, with the addition of think-aloud protocols, was therefore considered an appropriate combination of knowledge elicitation methods for this study.

Methodology

The knowledge elicitation method of process tracing with concurrent think-aloud verbal protocols was used while subjects completed a set modelling task (this task is described below). It is argued that this method is appropriate for collecting declarative command knowledge and specific procedural command knowledge. It also enables the identification, through both direct and inferential techniques, of strategic knowledge but not the reasoning processes underlying the procedural knowledge. The design of the task used in the study needed to meet a number of criteria. The task was sufficiently complex to challenge the expert while, at the same time, the modelling process would not take too long to complete. In this manner the expert would be engaged in, but would not feel compelled to devote excessive amounts of time to the process. The task incorporated a number of parts, both main and peripheral, so that parsing needed to occur and decisions regarding the category of part (main, peripheral) and the order of part generation could be made. The individual parts were capable of being modelled using a number of different algorithms so that choices of strategy would be made and opportunities existed for efficiencies of production to occur through a strategic approach. The designed model is illustrated in Figure 1 below.

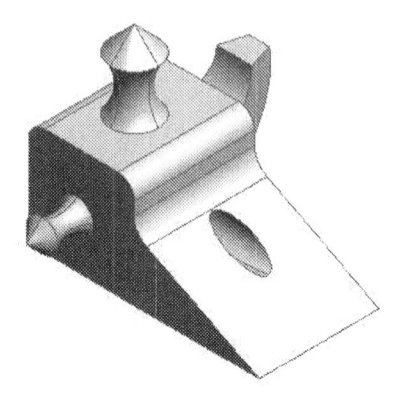

Figure 1. CAD Task

Data collection for the subjects was undertaken in their individual workplaces without the presence of the researcher. This procedure was chosen due to the disparate locations of the subjects and in order to overcome the possibility that the presence of a researcher may have inhibited the verbalisation of the subject. Subjects were provided with a package of materials including: instructions for the installation of the Camtasia software, drawings of the object to be modelled, CD ROM containing the software and onto which the data was to be saved and returned for analysis, a headset microphone for use when undertaking the process tracing and think-aloud protocols and a stamped addressed envelope for the return of the CD ROM.

During the construction of the model, data capture using video capture, with associated verbal protocols, was undertaken. Consistent with the findings of Ericsson and Simon (1993) specific instructions were given to the subjects to avoid the possibility that they would try to explain their reasoning. Subjects were instructed to "talk aloud constantly from the time you see the problem until you have generated the solution. I don't want you to plan what you say or try to explain what you are doing" (Ericsson & Simon, 1993, p. 376). Further, subjects were instructed to trial the use of the video capture software with their CAD system prior to actual data capture so that they would gain experience in the process of verbalising their thoughts while they undertook a CAD task. Subjects were also instructed to undertake the process of data collection alone. These procedures were incorporated into the instructions so that subjects became more comfortable with both the data capture and concurrent verbalisation processes. Analysis of the expert data gathered through this knowledge elicitation technique was designed to provide insight into the cognitive processes undertaken by CAD experts involved in the process of constructing 3D-CAD models. The process of data segmentation prior to analysis is described in the following section.

Data segmentation

Verbal protocols were fully transcribed and individual images representing the sequential development of the computer model were captured from the video. These data were placed into a table with the image placed in the cell adjacent to the verbal protocols. In this manner the verbal protocols relating to the particular image were identified. Data segmentation for the purposes of analysis of the data from the study was thus undertaken on the basis of changes in the perceptual or screen image. When a noteworthy change in the screen image was generated by the subject a new row was inserted into the table and the particular image inserted into cell three of the row. It is important to note that the images in the table are representative only of a dynamic on-screen process. The full verbal protocol was inserted in cell two of the row. Individual thoughts are represented as sentences, or clauses; pauses are represented as a series of periods. The analysis of the CAD and cognitive processes is included in cell 6 of the row and the coding for CAD and cognitive operations are inserted into cells four and five respectively. (An example of the data segmentation and coding for Subject 2 is illustrated below in Table 1). The individual coding is described in the following section.

Table 1. Example data segmentation and coding - Subject 2.

	Verbal protocol	CAD Image	Code 1		Analysis
			CAD	Cognitive	
	and let's trim it off. Now we need to trim off and select these as a smart line and it is continuous now. We need to delete the last little bit if we can no!! no!! no!! UNDO!!			MS CM	Mental subtraction is used to ascertain which sections of the line/circle need to be removed to get the correct profile. Following the deletion process the result is checked for accuracy.

IVAN CHESTER

	That I said drop ... I didn't say drop a line string! Delete that last little bit. Okay we've done that now let's revolve that. Oh bugger!!! I didn't join them up while I had the chance. Take that point and that one .... finished.		RS	CM MS E	Following checking the operation is undone and reconstructed correctly. Imaging of correct mental subtraction enables this process to be corrected. Rotation sweep enables the completed shape to be imaged prior to competing construction. Subject evaluates the process used.
	Let's revolve this little sucker yep axis of revolution yep yep yep yep done.		RS	CM	Subject uses rotation sweep in order to generate the desired solid from the produced profile, the direction and extent (number of 90° sectors) of the operation is imaged and constructed. Resultant construction is then checked.

For explanations of codes, see below.

Coding the protocols for CAD and cognitive processes

Data collected needs to be coded so that the number of the cognitive process used by the experts may be quantified, inferences drawn about the nature of the processes undertaken and comparisons made between experts. Coding of the data for this project was undertaken through an analysis of both the verbal protocol and the video images and a summary table for each subject was generated identifying a number of different categories of CAD processes and cognitive processes. These categories were identified on the basis of previous research in both cognitive science and the particular domain under investigation and are described in the following section.

CAD process categories

A number of separate CAD processes are used in three-dimensional solid CAD modelling and were therefore able to be identified within the data from experts engaged in the study. These CAD processes were identified through the algorithmic processes chosen by the subjects and through the on-screen image manipulations evident in the video.
- Multiple View [MV] - creating multiple views
- Rotation Object [RO] – rotating the CAD model to a new orientation.
- Rotation Sweep [RS] – creating a solid object by rotation of a cross-sectional shape around a central axis.
- Rotation on Path [RP] – creating a solid object when a cross-sectional shape is rotated or swept along a path or axis to produce, for example, tubular shaped objects.
- Rotation of Individual Surface [RIS]
- Construction Extrude [CE] – extruding an object
- Object Resizing [OS] – changing the size of an object.
- Construction Lofting [CL] – using a number of planar shapes to generate a solid object. This process is sometimes referred to as blending.
- Mirror Imaging [MI]
- Geometry Identification [GID] –identifying the geometry necessary to produce a given solid.
- Geometry Construction [GC] – constructing specific geometry.

Cognitive process categories

In this study it was hypothesised that the strategic knowledge of three-dimensional solid modelling CAD experts is characterised by the use of a range of mental imagery and metacognitive processes and these divisions were used as a basis for analysis. The following cognitive processes involving mental imagery were identified in the data:
- Mental Deconstruction (including parsing) [MD] (Verstijnen, van Leeuwen, Goldschmidt, Hamel, and Hennessey, 1998)

- Core Part Selection [CP] (Hoffman and Singh, 1997)
- Mental Construction [MC] (Cooper, 1990; Finke & Slayton, 1988; Kosslyn, Reiser, Farah & Fliegel, 1983)
- Mental Subtraction [MS] (Verstijnen, van Leeuwen, Goldschmidt, Hamel, and Hennessey, 1998)
- Spatial Positioning [SP]. (Cooper & Shepard, 1978; Lord, 1985; Shepard, 1974)
    The following metacognitive processes were identified in the data:
- Planning [PL] – planning or goal setting (Anderson, 1993)
- Strategy Selection [SS] (which CAD strategy or algorithm to use) (Bhavnani 2000; Bhavnani & John, 1996)
- Predicting Consequences [PC] (Gaughran, 2000; Rodriguez, Ridge, Dickinson & Whitwam, 1998)
- Checking / Monitoring [CM]
- Evaluation [E] (Veenman & Verheij, 2003; Resnick, 1976; Sternberg, 1990).

Analysis

The data for Subject 2 on the study is used below as an example. Subject 2 successfully completed most aspects of the modelling task in a time of 32 minutes 59 seconds. The 'horn' shape was unable to be modelled by this subject due to a combination of an inability to find an alternative algorithm in the software and unwillingness to spend the time necessary to calculate the size change for the transition that would have enabled the construction of the part using the algorithm with which the subject was familiar. Table 2 presents a summary of the CAD processes, cognitive processes involving mental imagery and metacognitive processes used by Subject 2.

RESEARCHING EXPERTISE DEVELOPMENT

Table 2 Subject Two – CAD and cognitive process totals

	CAD Processes			Cognitive processes involving mental imagery	
Code		No.	Code		No.
MV	Multiple View		MD	Mental Deconstruction - Parsing	9
RO	Rotation of Object	9	CP	Core Part Selection	1
RS	Rotation Sweep (Revolve)	5	MC	Mental Construction	9
RP	Rotation on Path		MS	Mental Subtraction	11
RIS	Rotation of Individual Surface		SP	Spatial Positioning	14
CE	Construction Extrude	3		Metacognitive Processes	
OS	Object Resizing	1	Code		No.
CL	Construction Lofting	1	PL	Planning	2
MI	Mirror Imaging	1	SS	Strategy Selection	1
GID	Geometry Identification		PC	Predicting Consequences	
GC	Geometry Construction		E	Evaluation	4
			CM	Checking/Monitoring	24

Analysis shows that Subject 2 engaged in a range of CAD processes during the construction of the model. These processes include those aspects of both declarative command knowledge and specific procedural command knowledge needed to undertake the specific task. In addition the two expected categories of cognitive processes were evident. Subject 2 engaged in a range of mental imagery supported processes specifically related to the controlled selection and execution of appropriate CAD algorithms. These processes include high instances of spatial positioning, mental subtraction, mental construction and mental deconstruction. These mental imagery processes were undertaken in order to visualise methods by which the model or the geometry could be constructed and to visualise the relationships between various parts of the model.

Accordingly, subject 2 utilised frequent checking and monitoring strategies, a number of which were associated with the CAD process of object rotation. This technique was used to enhance the process of checking and monitoring progress through re-orientation of the partially completed model in order to view it from a specific viewpoint. Subject 2 engaged in a number of evaluative processes thereby identifying when alternative modelling strategies would have been more efficient.

Discussion

The knowledge elicitation methodology of process tracing with concurrent verbal protocols using video capture software was found to be particularly effective in the acquisition of fine grained data. These data were able to identify both the declarative knowledge and specific procedural command knowledge of 3D-CAD experts. In addition, this methodology enabled the identification, both directly and through inference, of the use of a range of metacognitive processes used by 3D-CAD experts. These included the automatic parsing of objects into algorithmically salient parts and a range of visualisation techniques used to identify the process by which the object might be modelled (e.g. mental construction and mental subtraction) and checked for accuracy (e.g. mental rotation). The findings also demonstrate that the 3D-CAD experts, in a similar fashion to experts in other domains, made frequent use of a range of the metacognitive processes of planning, strategy selection, predicting, checking and monitoring. The remaining key issue identified through the study was the expert 3D-CAD strategy of selecting modelling approaches that used algorithmic commands such as fillet or round for model generation, in preference to geometry, in order to maintain model integrity and make the model easier to modify at a later time in the design process.

The methodology developed for this study will have application in research involving complex computer applications other than 3D-CAD. These could include applications such as robotics, computer control, web page development, systems analysis and graphic design. In order to gain additional data regarding the cognitive processes undertaken by subjects, it is possible that the technique could be modified and used to overcome the disadvantages of collecting think-aloud protocols which may interfere with the cognitive processing of the individual when undertaking complex tasks. For example this technique could involve video-capture of the initial modelling process and initial analysis by the researcher. Video-capture could again be used while the initial video-capture file is replayed to the subject in a process of stimulated recall while recording verbal protocols of explanations for the strategies that were used. The initial video could be paused to enable detailed explanations and individual aspects highlighted through mouse movements. The result would be a video of a video that included the verbal protocols of the stimulated recall. Final analysis by the researcher could then follow with the original video and the stimulated recall data in the one file. Further discussion regarding the efficacy of stimulated recall as a research methodology is found in Chapter 11.

REFERENCES

Anderson, J. R. (1993). Problem solving and learning. *American Psychologist*, 48, 35-44.

Bhavnani, S. & John, B.E. (1997) From sufficient to efficient usage: An analysis of strategic knowledge. *In Conference Proceedings Chi 97*. USA: AVM Press.

Bhavnani, S. (2000). Designs conducive to the use of efficient strategies. In Conference Proceedings *Symposium on Designing Interactive Systems*. USA: ACM Press.

Charness, N. & Schultetus, R.S. (1999). Knowledge and expertise. In Durso, F.T. (Ed.) (1999) *Handbook of applied cognition.* West Sussex: Wiley and Sons.
Charness, N. & Schultetus, R.S. (1999). Knowledge and expertise. In Durso, F.T. (Ed.) (1999) *Handbook of applied cognition.* West Sussex: Wiley and Sons.
Cooke, N.J. (2000) Knowledge elicitation. In Durso, F.T. Ed (2000) *Handbook of applied cognition.* West Sussex: Wiley and Sons.
Cooper, L.A. (1990). Mental representation of three-dimensional objects in visual problem solving and recognition. *Journal of Experimental Psychology: Learning, Memory and Cognition, 16,* 1097-1106.
Cooper, L. and Shepard, R. (1978). Transformations on representations of objects in space. In Carterette, E. and Friedman, M. (Eds.) *Handbook of perception.* Vol. 8, Academic Press.
Ericsson, K.A. & Crutcher, R.J. (1991). Introspection and verbal reports on cognitive processes – Two approaches to the study of thinking: A response to Howe. *New Ideas in Psychology,* 9, No 1, 55-71.
Ericsson, K.A., & Simon, H.A. (1993). *Protocol analysis: Verbal reports as data.*
Evans, G. (1991). *Learning and teaching cognitive skills.* Hawthorn: Australian Council for Educational Research.
Finke, R.A., & Slayton, K. 1988. Explorations of creative visual synthesis in mental imagery. *Memory and Cognition, 16,* pp. 252-257.
Gaughran, W.F. (2000). Modelling and design intent. *Engineering and Product Design Education International Conference,* pp. 61-68. Sussex, England.
Gordon, S. E. (1992). Implications of cognitive theory for knowledge acquisition. In R. R. Hoffman (Ed.), *The psychology of expertise: Cognitive research and empirical AI,* pp. 99-120. New York: Springer Verlag.
Hoffman, R.R., Shadbolt, N.R., Burton, A.M. & Klein, G. (1995). Eliciting knowledge from experts: A methodological analysis. *Organizational behavior and human decision processes, 2*(2), 129-158.
Hoffman, R.R., Shadbolt, N.R., Burton, A.M. & Klein, G. (1995). Eliciting knowledge from experts: A methodological analysis. *Organizational behavior and human decision processes, 62*(2), 129-158.
Hoffman, D.D., & Singh, M. (1997). Salience of visual parts. *Cognition, 63,* 29-78.
Kosslyn, S.M., Reiser, B.J., Farah, M.J. & Fliegel, S.L. (1983) Generating visual images: Units and relations. *Journal of Experimental Psychology: General, 112,* 278-303.
Lord, T. (1985). Enhancing visuo-spatial aptitude of students. *Journal of Research in Science Teaching, 22,* 395–405.
Matlin, M.W. (2005*). Cognition.* New Jersey: John Wiley and Sons.
Mieg, H.A. (2001). *The social psychology of expertise: Case studies in research, professional domains, and expert roles.* Mahwah, NJ/London: Lawrence Erlbaum Associates.
Resnick. L.B. (1976*). The nature of intelligence.* New Jersey: Lawrence Erlbaum Associates, Hillsdale.
Rodriguez, J., Ridge, J., Dickinson, A. and Whitwam, R. (1998). CAD Training Using Interactive Computer Sessions. *Proceedings of the 1998 American Society for Engineering Education Annual Conference & Exposition.* Retrieved 28/8/03 from
http://www.asee.org/conferences/search/00048.PDF
Rowe, H.A.H. (1985). *Problem solving and intelligence.* Hawthorn, Victoria: Australian Council for Educational Research.
Schraagen, J.M., Chipman, S.F., and Shalin, V.L. (1999). *Cognitive task analysis.* London: LEA.
Shepard, R.N. (1974). Representation of structure in similarity data - Problems and prospects. *Psychometrika, 39,* 373-422.
Sternberg, R.J. & Grigorenko, E. (2003). *The psychology of abilities, competencies, and expertise.* New York: Cambridge University Press.
Sternberg, R.J. (1990). *Metaphors of mind; conceptions of the nature of intelligence.* Cambridge, New York: Cambridge University Press.
Veenman, M.V.J. & Verheij, J. (2003). Technical student's metacognitive skills: relating general vs. specific metacognitive skills to study success. *Learning and Individual Differences, 3,* 259-272.

Verstijnen, M, van Leeuwen, C., Goldschmidt, G., Hamel, R., and Hennessey, J.M. (1998). Creative discovery in imager and perception: Combining is relatively easy, restructuring takes a sketch. *Acta Psychologica*, *99*(2), 177-200.

Ivan Chester
Faculty of Education
Griffith University
Australia

MARILYN FLEER

A CULTURAL-HISTORICAL PERSPECTIVE ON RESEARCH IN DESIGN AND TECHNOLOGY EDUCATION

INTRODUCTION

Over the past five years researchers have struggled to move thinking forward in technology education. Constructivist theory has provided researchers and practitioners with important directions, but new research methodologies are needed if technology education is to move forward. In drawing upon Vygotsky's (1987) writings, this chapter introduces a cultural-historical perspective on research in technology education research.

Vygotsky (1987) brought to our attention the importance of identifying the core concepts inherent in technology knowledge, and combining these with children's personal everyday knowledge. Vygotsky (1987) argued that much of what we know about concept formation has been derived from research methods which focus on the 'completed process of concept formation with the ready-made product of that process' (p. 121). Vygotsky (1987) argued that the dynamics of concept formation - how it develops, how it begins and what it looks like at the end – have not been systematically examined. He suggested that when we study the child's definitions of a particular concept, we are studying his (sic) knowledge or experience and the level of his verbal development more than we are studying his thinking in the true sense of the word' (p. 121). Studying the dynamic processes, as opposed to the 'end product', offers a useful research framework for technology education. This approach is particularly pertinent for researchers interested in how very young children pay attention to, and grow their understandings of, technological concepts.

This chapter exemplifies cultural-historical research in action through detailing a study which captured the dynamic nature of concept formation over time. In particular, the study examined recent thinking in 'ideas technology', and sought to make visible the dialectical relations between children's everyday concepts and schooled technology concepts. Hedegaard and Chaiklin (2005) have advanced the theoretical idea of a double move approach, which specifically emphasizes the 'relations among children's already acquired everyday concepts, subject-matter concepts, and local knowledge' (p. 69). A teacher who uses a double move, considers the child's everyday concepts (one move) and the academic school concepts (second move) at the same time. The double move approach operationalises how educators can bring together technology concepts with children's personal knowledge and their everyday practices. Researching teacher thinking alongside of children's technological activity requires sophisticated research tools and new ways of analysing data. The particular 'moves' the teacher makes will be discussed and illustrated through data sets of children and teachers working technologically in their classroom.

These theoretical and methodological ideas provide new directions for researchers as they grapple with institutional and societal demands for increasing outcomes for children in technology education. These new directions provide both promise and challenge, as noted by Davydov and Zinchenko (1993) who argue that a researcher:

> ...must pay for (such) an investigation of the mind with his (sic) sweat; he must seek a theory that will help him to understand the human mind and consciousness (Davydov and Zinchenko, 1993: 98).

CULTURAL-HISTORICAL PERSPECTIVE ON RESEARCH

Cultural-historical theory has provided practitioners and researchers with a new framework for investigating teaching and learning in technology education. A cultural-historical view emphasizes the relationship between individuals and the broader social and technological world. Rather than focuses primarily upon what an individual knows and can do, as is featured in a constructivist orientation to researching learning, a cultural-historical worldview goes beyond what is in the 'head' and examines the social context in which individuals reside. As Chaiklin (2001) suggests, cultural-historical research is 'the study of the development of psychological functions through social participation in societally-organised practices' (p. 21). This orientation foregrounds different data, as historical and societal dimensions must be considered alongside of data generated by mapping individual constructions of knowledge. As Wertsch (1998) suggests:

> Methodological individualism assumes that cultural, institutional, and historical settings can be explained by appealing to properties of individuals, and social reductionism assumes that individuals can be understood only be appealing to social fact (p. 179).

A cultural-historical perspective on research seeks to use a unit of analysis that focuses on the dynamic relations between settings and individual mental processes (Wertsch, 1998: 179).

A cultural-historical view of research privileges how individuals and groups of people interact within their social and technological environments. Learners are shaped by their community, and they in turn change their community through their participation. Wertsch (1998) suggests that the cultural tools available to us in a community are the product of centuries of social relations and artefact creation. Each time we interact with our designed environment (whether an artefact or a social rule) we appropriate cultural-historical contexts and processes which determine what is possible and not possible. Whenever we interact in our designed world, we engage with an existing set of technologies and possibilities which shape how we think and interact. A cultural-historical perspective suggests that individuals do not operate on their own, but rather are part of a cultural-historical context, and these relationships and contexts must feature in the research design. For instance the QWERTY keyboard still exists within our computer designs, not because the particular configuration of keys are needed to ensure mechanical smoothness when pressing buttons (as was the case for typewriters), but because the keyboard design has been institutionalised within society. History leaves its mark not only in contemporary artefacts but also social relations. Wertsch (1998) writes that everyday practice can only be understood if we take a broader cultural-historical perspective of everyday action. He suggests that 'cultural tools are historically situated, and this history typically leaves its traces on mediational means and hence on mediated action' (p. 63). The research lens must broaden to take account of these cultural and historical factors when examining everyday individual action in our technological world.

Chaiklin (2001) argues that cultural-historical psychology, and therefore cultural-historical research, is institutionally quite young. He states that 'if judged by the calendar and the historical span of empirical psychology, then cultural-historical psychology – as reflected in the research traditions that have developed from Vygotsky, Leontiev, Luria, and their colleagues – is relatively old' (p. 15). However, Chaiklin (2001) argues that cultural-historical psychology 'does not have coherent institutional structures normally associated with a research tradition, thereby making it more difficult to define and locate the tradition in a singular manner' (p. 15). This is particularly pronounced in the area of technology education research, where few scholars have applied a cultural-historical approach to their studies. This is an interesting phenomenon, given that technological activity exemplifies the relations between individuals and their social and technological world. Like cultural-historical psychology, technology education represents an old tradition of learning. However, in the context of primary education, it is a relatively new area of learning, with a very brief research tradition. Scholars from science education, steeped in constructivist theory have broadened their research to encompass technology education. Few scholars of 'primary technology education' have come from within the ranks of primary technology education classrooms (a notable exception is Stables), or secondary technology education classrooms (notable exceptions are Seeman, Wellbourne-Wood, Williams). As such, the application of cultural-historical theory to research in technology and design education, is still in its infancy.

THE STUDY OF CONCEPT FORMATION – HOW CAN IT INFORM TECHNOLOGY EDUCATION?

'I do not want to discover what the mind is on the cheap, by culling a few quotations,' wrote Vygotsky. 'I want to learn, by Marx's formal method, how a science (research approach) is constructed and how to approach the study of the mind' (1982-84, vol. 1: 421) (Davydov and Zinchenko, 1993: 98).

Contrary to research in concept formation at the time, Vygotsky (1987) undertook an experimental study which sought to develop an understanding of children's concepts in action. He argued that research had concentrated upon identifying children's concepts at the end of the conceptual process. This he suggested, was a static process of mapping mature concepts, and did not help explain how concepts evolve and change over time. He argued that for a child to express a perspective during data gathering, he or she would already have developed conceptual understandings about what was being researched. This type of research would not show children's thinking, but would rather present a 'ready made' concept, already mature and fully formed. Vygotsky (1987) argued that it was more fruitful to examine the development of a concept. He suggested that traditional research focused almost exclusively on 'the word'. "It overlooks the fact that, for the child in particular, the concept is linked with sensual material, the perception and transformation of which gives rise to the concept itself" (p. 121). According to Vygotsky (1987), it is both the word and the physical (sensual) act, which require investigation. In technology education, both the word and the physical act are required by children when learning. Learning occurs with both the hands and the minds – each is dependent upon the other. This symbiotic relationship has been the focus of much philosophical discussion (see Scharff & Dusek, 2003), but it has not received the depth of classroom research attention required to understand concept formation in technology education fully (see Moreland et al., 2001).

Through experimentation with blocks, where the researcher progressively included additional conceptual information about how to sort the blocks as the child worked (so as to capture the dynamic nature of the activity), Vygtosky (1987) identified the dialectical relations between everyday and academic concept formation. Through this research approach, Vygotsky (1987) found a way of capturing the dynamics of concept formation in everyday practice, and academic or schooled concept formation. Knowing more about the relationship between everyday practice and important technology concepts learnt at school, is important for understanding how primary children build concepts. This becomes particularly important when we consider recent develops in technology education, where ideas rather than simply artefacts are featured. Societies are increasingly trading in ideas rather than manufactured goods. As a global community, our children – the future designers and decision makers – require different conceptual tools. Technology education provides an important vehicle for developing these conceptual tools. However, little research has been directed towards ideas technology, and few researchers have considered the implications of these new research questions when drawing upon research methodologies. Research approaches which capture the conceptual relationship between thinking and artefacts, provide a way forward for research which seeks to concentrate on "ideas technology". Finding research tools which better understand how children's theoretical ideas develop is more likely to inform contemporary approaches to technology and design education.

Davydov (1988) suggests that theoretical knowledge is built from the interaction of people with societal problems and contexts (see Hedegaard, 1995). Knowledge generation is embedded within its context. Empirical knowledge is disembedded from context. For instance, "ideas technology" foregrounds culture, ethics, values and creativity. What kinds of research tools are needed to examine these theoretical ideas? What kinds of concepts develop when "ideas technology" are taught, and how can we research contemporary approaches to technology and design education?

CULTURAL-HISTORICAL TOOLS FOR MAPING NEW APROACHES TO TECHNOLOGY AND DESIGN EDUCATION

A cultural-historical methodology framed a study which sought to examine whether or not the new vision of technology education could be operationalised within the early years of school. (see Technology education – A shared vision; Department of Education, Science and Training, 2002). In particular, the study investigated how children would engage with the concept of "ideas technology" and examined the types of thinking that would emerge amongst the children. In this section, a cultural-historical methodology is detailed. Within the context of this study, the data gathering technique will be illustrated.

Research methods which foreground group learning

In keeping with a cultural-historical view of research, the study investigated group thinking within a classroom over an extended period of time (four weeks). Rather than focussing on individual learning, the camera lens roamed to capture group interactions, group constructions of ideas, and group technological activity.

Figure 1: Focusing on 'group thinking'

RICHARD KIMBELL

Data gathering concentrated upon how groups of children framed their ideas and how their comments or contributions influenced each other. The research principle that drove the data gathering was that learning is more than a construction by the individual. In a group discussion, children are working with ideas and concepts, and collectively their comments and thinking are stronger than if the researchers approached them individually to 'deconstruct their learning' as though it were occurring in isolation of the social practices and collectivity of thinking. Traditionally the research lens has examined the efforts of individuals, what they say and do in technology. A cultural-historical perspective examines the social spaces between individuals, rather than the individuals themselves.

Research methods which make visible historical traces of design and invention

A cultural-historical perspective on research foregrounds the importance of the historical dimensions of context. This view is more than simply considering the historical dimensions of the artefacts that children may be using or constructing, but rather it is a recognition that previous inventions and contemporary inventions are inextricably linked to each other. For instance, Wertsch (1998) argues that in Western communities, patents and copyright laws have been invented to protect individual ingenuity and invention. However, an invention is never the sole act of an individual. As a community, we generate societal needs and create rules, systems, and artefacts, and have shared values and ethical practices – all of which shape how an individual thinks and acts. The task of inventing must be considered to be located within the 'design world of materials', 'the design world of tools' and within the context of community values and beliefs. An individual never acts on their own. These historical traces shape how an individual designs and how the design evolves – every act represents a history of inventions and designs.

Figure 2: Methodologies which capture historical traces

In the study reported in this chapter, the researcher documented how children interacted with the resources and captured on tape, how the children talked about, and personalized the resources. Rather than ignoring their play conversations with the materials or regarding their comments as irrelevant, they were examined closely to see how they connected with their everyday lives and experiences.

Research methods which foreground the interface between everyday concepts and academic concepts

Cultural-historical research builds links across contexts. For young children, learning is not confined to classroom contexts. Indeed cultural-historical research views concept formation occurring when the everyday practices of children are fused with schooled academic knowledge. In the study reported in this chapter, the researchers were interested in how ideas and concepts introduced in the classroom were linked to the children's home world. As such, the children created 'ideas books' which they used to note their designerly thinking, but also to map everyday practices at home which linked to what they were investigating at school. The children developed interview questions, and set about phoning each other up over the weekend in order to find out more about each others' everyday contexts and designed environments.

Figure 3: Methodologies which move across home and classroom contexts

The children's thinking books, their drawings, their interviews of each other, and their group posters, all provided a rich data set for analysis. Analysis of the interlacing of everyday concepts and academic technology and design concepts was an important dimension of the research.

In qualitative research of this kind, analysis is complicated as it requires the researcher to identify patterns which are changing during the course of the data gathering period. For instance, Rogoff (2003) suggests that snapshots of data can be analysed, but that they should be conceptualised within an analysis which views the data like a filmstrip of moving images over time and space. Each snap shot, is like a still frame within the filmstrip (or a photo frame from within a digital video recording, as occurred within this particular study). Rogoff (2003) uses three lenses to analyse each of these snap shots of data: Individual plane of analysis (1st lens), Interpersonal plane of analysis (2nd lens), and a community/institutional plane of analysis (3rd lens). In the study discussed in this chapter, data which focused on what individuals expressed or said was analysed using the first lens. How the children interacted and influenced each other was examined using the second lens, and finally, the way the classroom was organised or how the resources influenced learning, was examined using the third lens. Taken together, these lenses allow for the whole context – children, teacher, artefacts, teaching program, classroom and home context to be examined concurrently. However, the three lenses allow for a more systematic analysis of the data, since particular aspects of the data set could be foregrounding with a particular lens, whilst the whole context was put into the background (but not lost). Being systematic with the analysis ensured that a large body of 'moving and changing data' could be managed and examined more robustly.

Researching the double move in teaching

In the context of mapping the conceptual connections between everyday home practices, and the concepts being explored in the classroom, it is possible to caste the research lens onto the teacher. We can ask, what is her focus of attention? Is the teacher explicitly working on fusing children's already acquired everyday concepts, and academic technology concepts within the context of local practice and knowledge? Researchers following a cultural-historical methodology, examine how the teacher is framing experiences for the children. Does the teacher simply focus on the academic concepts or does the teacher also put in place structures and processes for finding out children's everyday concepts and experiences? Through examining the interlacing structures put in place by the teacher, it is possible to note whether or not the teacher uses a double move in teaching. In the study reported in this chapter, the researcher created an 'evolving PowerPoint' that captured photos, scanned work samples, video clips, and transcribed comments, over the period of the data gathering period. On the second data gathering trip, the children were shown the PowerPoint which was located on a lap top in the corner of the room. The children were invited to go through the PowerPoint whenever they wished over the course of the study. The PowerPoint was updated after each data gathering visit. The children were encouraged to talk about the slides, and their responses were also documented. The evolving PowerPoint represented the community perspective that was being privileged in the classroom and in the research method. The teacher also discussed the slides with the children and with the researcher. These interactions with the data provided a vehicle through which the researcher could document all comments which related to the teacher's explicit consideration of the everyday concepts and academic technology concepts.

Figure 4: Documenting the double move

SUMMARY

Cultural-historical theory provides a powerful theoretical orientation for research in design and technology education. Taking account of the context, which includes the historical and cultural dimensions of the technologies being used or studied, has been under researched in design and technology education. This kind of research methodology primes us to look at data gathered in quite a new way and suggests new study designs not previously formulated. Rather than noting the individual and their thinking, we begin to pay attention to:
− how a collective of children interact and move ideas about in a group
− how concept formation may not simply be a product of an individual, but can be held concurrently across individuals

- how technological artefacts are socially shaped through history and by people, suggesting ways of working, thinking and designing with information, systems and materials – influencing what is possible
- how conceptual development moves and changes over time and space
- how social relations shape design and technology activities

In cultural-historical research, we focus our attention on quite different dimensions of design and technology pedagogy and the associated cognition of children. It is likely that this approach to design and technology research will yield new discoveries in pedagogy and cognition. As such, cultural-historical theory offers a new way forward for design and technology education research.

REFERENCES/BIBLIOGRAPHY

Chaiklin, S., (2001) The institutionalisation of cultural-historical psychology as a multinational practice. In Chaiklin, S., (ed.). *The theory and practice of cultural-historical psychology*. Aarhus University Press, Denmark.

Davydov, V.V., (1988) Problems of developmental teaching. *Soviet Education, 30* (8), 6-97.

Davydov, V. V. and Zinchenko, V. P., (1993) Vygotsky's contribution to the development of psychology, In Daniels, H., (ed.) *Charting the agenda. Educational actvitiy after Vygosky*. Routledge, UK (pp. 93-106).

Department of Education, Science and Training., (2002) *Technology Education – A shared vision*. A workshop held in Melbourne, Victoria. (8-9th July, 2002: Hilton International Airport)

Hedegaard, M., (1995) The qualitative analysis of the development of a child's theoretical knowledge and thinking. In Martin, L.M. W., Nelson, K., and Tobach, E., (eds.) *Sociocultural psychology. Theory and practice of doing and knowing*. Cambridge University Press: USA (pp. 293-325).

Hedegaard, M. and Chaiklin, S. (2005) *Radical-local teaching and learning A cultural-historical approach*. Denmark, Aarhus University Press.

Moreland, J., Jones, A., Milne, L., Chambers, M., and Forret, M., (2001) An analytical framework for describing student learning in technology. Paper presented at the 32nd ASERA Conference, 11-14 July 2001, Sydney, Australia

Rogoff, B., (2003) *The cultural nature of development*. Oxford University Press: Oxford UK.

Scharff, R. C., and Dusek, V., (2003) (eds.) *Philosophy of technology. The technological condition. Aan anthology*. Blackwell Publishing: Australia.

Vygotsky, L.S. (1987) Thinking and speech. In L.S. Vygotsky, *The collected works of L.S. Vygotsky, Vol. 1, Problems of general psychology*. (pp.39-285). R.W. Rieber & A.S. Carton (eds); N. Minick (Trans.) New York: Plenum Press.

Wertsch, J. V., (1998) *Mind as action*. Oxford University Press: USA.

Marilyn Fleer
Faculty of Education
Monash University, Peninsula Campus
Australia

RICHARD KIMBELL

DESIGN PERFORMANCE: DIGITAL TOOLS: RESEARCH PROCESSES

INTRODUCTION

This chapter will begin with a summary review of the research processes that we have used in the Technology Education Research Unit (TERU) at Goldsmiths College to promote learners' design performance for the purposes of assessment. Specifically I will outline the approaches initially developed in the 1980s for the APU national survey of design & technology capability and how these evolved through a number of research projects (in the UK and internationally) for broader evaluative purposes.

Using the framework established through this review, the bulk of the chapter will be devoted to a description and analysis of the transformations to these research processes that have been necessitated in our latest project *e-scape*. The distinctive feature of this project lies in the use by learners of digital tools for design performance and their resulting design portfolio emerging *virtually* in a secure web-space.

The technologies involved in learners' design performance have dramatically shaped the processes of data capture for research processes. As but one example, for the 1st time we can use, in real-time, the authentic *voice* of learners reflecting (through a series of 'sound-bites') on their emerging prototypes. These new data capture opportunities to transform the kinds of questions that we can ask but equally they extend the methodological challenges in the research.

The chapter will draw from data arising from the first national pilot study of the *e-scape* system (June/July '06) in 14 schools across England. Uniquely, the resulting performance from 300+ learners is simultaneously available – through the web-space – to researchers across the globe. That alone poses some interesting research questions.

PERFORMANCE BASED ASSESSMENT

The first project undertaken by TERU was commissioned by the Assessment of Performance Unit (APU), which was a research branch of the Department of Education & Science (DES). Established in 1975, the APU surveyed and monitored national levels of achievement in schools, and in the first decade of its existence it adopted a dominantly scientific paradigm – with 'tests' delivered systematically to a randomly selected 2% sample of any chosen age/group or population. Extensive surveys were undertaken in mathematics, English, science and modern languages, typically at ages 8, 11 and 15 and much had been discovered about what learners could be expected to achieve in these subjects at those ages. As the APU evolved, however, it gradually acquired a purpose beyond testing & measurement and increasingly came to see itself as providing support for curriculum development.

> Early APU surveys were seen largely as providing data about what learners could or could not do - and how this changed over time. In curricular terms APU was distinctly non-interventionist. Progressively, however, the concern became to understand why learners performed in the ways they did; teasing out learning blocks and helping teachers to enhance learning. APU was increasingly becoming a force for curriculum development. (Kimbell et al., 1991 p. 11)

Our work for APU (1985-91) fitted neatly into this more interventionist/ curriculum support view of research, and probably the most significant element of that work concerned the development of 90 minute capability-based tests of learners' performance in design & technology. The essence of our approach was to create design activities within which learners' responses were structured through two interacting devices; a *script* that required the teacher/administrator to introduce elements of the task in carefully structured sequences of sub-tasks, and a response *booklet* that enabled the learner to operate on those sub-tasks whilst keeping in mind (and in vision) their work on the immediately pre-ceding sub-tasks.

These are not trivial devices – but evolved through a series of explorations and trials through 1985, '86 and '87. The generic structure – which sought to create a response mode in learners that iterated between dynamic/generative and critical/reflective – was developed into 24 different activities set in 6 contrasted contexts. The final form of these 90 minute activities emerged in 1988 as the APU design & technology survey in which we 'tested' the design & technology capability of 10,000 learners in 700 schools through England, Wales and N Ireland. A full description of this work was first available in Kimbell et al (1991) and has subsequently been presented as part of the evolving TERU story in Kimbell and Stables (2006).

It would be fair to say that when the tests emerged in the late 1980s they were nothing like the conventional concept of what a 'design project' was supposed to look like. The teacher had no means for controlling (or even steering) learners' ideas. They could read out the administrator script, but otherwise were powerless to intervene in learners' work. It was therefore something of a surprise to many of these teachers that the resulting work from learners was so powerful.

TWO KINDS OF CONTROL

What had emerged from our explorations was the realisation that two very different kinds of control can be exerted by teachers in a design project.
procedural control   *(what to do* next*)*
idea control   *(the* substance *of the idea)*

An example of the former might be manifested in a comment like "have you thought about how you might test that out?" An example of the latter might be represented in a comment like "won't you need some kind of switching system?" Our school-based explorations led us to the view that teachers were far more used to (and more comfortable with) steering the *content* of learners' ideas than they were the process. There were many reasons for this, but principal amongst them was the idea that teachers had a body of *content* to impart - about materials, systems, manufacturing and the rest. In these content terms, the teacher was the expert and the learner the novice, and teachers therefore felt the need to guide their charges through a tightly controlled series of 'projects' so that all the necessary 'stuff' outlined in the syllabus/scheme of work could be taught.

We might go so far as to say that projects were not really design projects at all in the sense that original or innovative ideas/products might emerge from them.

When the teacher said ...

"We are going to do a project on watering plants"

What s/he really meant was ...

"I'm going to show you how to make a circuit that will switch on and off in relation to the moisture in contact with two terminals. And you can all copy what I show you"

Wow – how exciting is that? In reality, such projects were a sham - providing a fig leaf of motivational purpose to an otherwise nakedly instructional programme of work. Little wonder that all the moisture sensors looked alike and behaved in exactly the same way. That was the point of it all. In fact – typically - the only ones that varied from this desired plan were the ones that didn't work at all, because the circuit failed in some way. This had the effect of further cementing in learners' consciousness the idea that they had better get better at copying the teacher's model, because difference (too often) = failure.

So, imagine the thought process of the teachers when (in 1988) design tasks were set to learners in an APU national survey of performance. The teachers naturally felt represented by their learners and wanted them to do well. Imagine the consternation when they found that the only 'steer' they could give to the selected learner group had nothing to do with the content of their design solution. They couldn't advise on materials, or systems or manufacturing processes – or anything. All they were allowed to do was read the script. And the only things that the script dealt with were *procedural* things like

We would like you now to think about the <u>user</u> ... Who is it for, and what will your product have to do to be successful?

I should be clear about why we were taking this strictly procedural view of things. The description of design & technology capability that we had articulated was essentially procedural, involving learners in taking on a task and then working it out from their earliest hazy ideas through a series of iterations that progressively enabled them to detail what it would be like: how it would work and be used. We were commissioned by APU to assess learners' performance, and – for us – that meant assessing their ability to do these things for real as they undertook a task. We therefore put in these procedural requirements to ensure that there would be evidence of learners' ability (or not) to do these things. The motive for us was the accumulation of evidence for assessment. If we didn't ask learners to tell us what they knew about or thought about their <u>users,</u> how could we assess their user awareness? So we built the requirement into the activity and could be confident that there would be some evidence for us to examine as a result.

AN UNEXPECTED CONSEQUENCE

But we were somewhat taken aback by the outcome of this strategy. It was apparent that we had evolved an approach that was not only very different from the prevailing project-based methodology of teaching but was also very effective in *promoting* performance. Our choreographed activities were designed to promote *evidence* of learners' performance but wherever we tried them we discovered that they also had the effect of promoting the performance itself. Our trials in schools had been prefaced on the idea that the activities had to be made to work effectively with learners, but it was not until teachers started to express their surprise about the performance of individuals that we realised that in the processes of assessing performance, we were enhancing it.

In a recent development of the approach for the Department for Education & Skills (DfES) and the Qualifications and Curriculum Authority (QCA) a learner summarised this phenomenon very succinctly:

Learner AB324: I realised that I could do more than I thought I could (Kimbell et al 2004 pp 36)

Throughout the APU project we received innumerable comments of this kind and they encouraged us to see the activities as more than just an APU assessment tool. We realized that they are better seen as an *approach* rather than a product. Not so much a test, more a way of working. And having *supported* learners' performance through the activity, they inevitably made it more *apparent* and therefore more available for assessment.

Having come to this realization, we were subsequently able to modify the approach for other projects where we were seeking either to support or assess learners' performance. As examples, we used the approach for the *Decisions by Design* project (Kimbell, R., Mahony, P., Miller, S., & Saxton, J. 1997) and for the *South Africa curriculum evaluation project* (Kimbell & Stables 1999).

By the time of the Design & Technology Millennium Conference in London we had created a whole series of variants on the approach with different time scales, different activity focus, and with different purposes, but all based on the idea of choreographed, structured activity. We wrote about the approach for that conference – describing the approach as *the un-pickled portfolio*. (Stables & Kimbell 2000). The title refers to the fact that whilst 'normal' design portfolios grow over an extended period of time and involve the learner becoming 'pickled' in the juices of the task, our approach is far too quick-fire to allow such pickling. And the portfolio that results is fresh and 'un-pickled'.

PROJECT E-SCAPE

A number of events and forces led to the creation of the e-scape project. In 1999, the then Department for Education & Skills produced the latest version of the National Curriculum (DfES 1999), and one of the welcome additions was the articulation of 'importance' statements, in which the vision of subjects is encapsulated. The Statement for design & technology made great play of the need for creativity, innovation and teamwork. Prest (2002), however, identified a misfit between these bold statements of intent and the reality of assessment processes in design & technology. He argued that our approaches were always focused on individuals (not teams) and tended to reinforce 'safe' rather than risky behaviour in learners; stifling rather than supporting design innovation. Discussions of these issues within DfES and the Qualifications and Curriculum Authority (QCA) led to the establishment of 'assessing design innovation' a research & development project in which we in TERU were commissioned to find ways to deal with assessment whilst specifically recognising the twin challenges of teamwork and innovation. The project ran between Jan 2003 and Dec 2004 and resulted in a model of assessment that was developed from the 'unpickled portfolio' concept and that emerged as a 6 hr structured designing/modelling activity. Subsequently the Oxford, Cambridge and RSA examination board (OCR) has adopted the approach for part of the assessment of its new Product Design GCSE course. It is presented there as one of the assessed modules – the innovation challenge. (OCR 2005)

Whilst *assessing design innovation* was the starting point for e-scape, there were other educational and political forces at work, specifically concerning e-learning. E-learning is a term that has emerged to describe a wide range of digitally enhanced educational experiences, and an overarching policy direction emerged from the Prime Minister's Strategy Unit. In a document entitled "Connecting the UK: the Digital Strategy", action 1 is defined as "Transforming Learning with ICT" and describes the need for everyone to have an electronic portfolio for lifelong learning:

> .. in the future it will be more than simply a storage space - a digital site that is personalised, that remembers what the learner is interested in and suggests relevant web sites, or alerts them to courses and learning opportunities that fit their needs. We will encourage all organisations to support a personal online learning space for their learners that can develop eventually into an electronic portfolio for lifelong learning. (Prime Minister's Strategy Unit, 2005, 30)

Developing a similar theme, the DfES e-learning strategy identified the provision of a centralised e-portfolio as an important priority for reform, with a personal identifier for each learner, so that education organisations could support an individual's progression more effectively. Together, these facilities were seen to lead to the creation of electronic portfolios, making it simpler for learners to build their record of achievement throughout their lifelong learning (DFES e-strategy 2005).

Within this climate of ICT developments in schools, it was noted that design & technology featured strongly on the radar, summarised in the OFSTED report of 2004.

Secondary design and technology (D&T) departments continue to make widespread and effective use of ICT in their teaching (OFSTED, 2004, para 121).

Following detailed discussions between DfES, QCA, GCSE Awarding Bodies and TERU, we were presented with the following specification for a research venture that we named e-scape (e-solutions for creative assessment in portfolio environments).

QCA intends now to initiate the development of an innovative portfolio-based (or extended task) approach to assessing Design & Technology at GCSE. This will use digital technology extensively, both to capture the student's work and for grading purposes (QCA Specification, June 2004, 1).

Essentially the e-scape system was to involve learners tackling design & technology tasks - using predominantly digital tools – such that their resulting work appeared not in a paper-based portfolio but rather in a website. The task was described as having four dimensions for exploration; technological, pedagogic, manageability, and functionality. Work on the project commenced in Jan 2005 and is due to be completed in Jan 2007.

METHODOLOGICAL CHALLENGES

There are many elements of this project that raise interesting methodological discussions and in this chapter I have chosen to focus on just three of them.
- the *concept of an e-portfolio* and the consequences of our decisions on this
- the learners' *voice* in the portfolio
- the challenge of *making judgements* on the quality of learners' portfolios.

CONCEPTS OF E-PORTFOLIO

The concept of 'portfolio' is problematic, arising in part from the fact that the term portfolio means very different things to different people. The potential for different interpretations is increased by the use of portfolios as an assessment tool, and complicated yet further in the context of e-learning, where 'e-portfolio assessment' has become a minefield of misunderstanding and confusion. In collaboration with our project LEAs and schools – inherited initially from assessing design innovation – we observed several forms of what a portfolio might be.

i. The most common form of 'portfolio' was as something akin to a box-file into which the learner (or perhaps the learner's teacher) places work to demonstrate that certain operations, or skills, or processes have been experienced. Viewed in assessment terms, the learner's portfolio becomes a collection of evidence that is then judged against some rubric to arrive at a mark or a level. A portfolio of this kind is conceived of as little more than a *container* for evidence. Translated into the *e-portfolio* world, it is possible to conceive of many ways in which the evidence being 'contained' (in a great big digital bucket) could be enhanced through the application of digital systems.

ii. A somewhat more sophisticated view of portfolio arises from process-rich areas of the curriculum, where teachers encourage students to document the story of a developing project or experience. This results in learners *reporting* what they have done at various points in the process. In this kind of 'presenting' or 'reporting' e-portfolio, it is not unusual for students to use linear digital presentation technologies – eg powerpoint – to give a blow by blow account of where they have been in the project – and how they finally got to the end.

Neither of these versions of e-portfolio captures the dynamic capability dimension that informs our view of a design & technology portfolio. The central problem – in both cases – is that the portfolio construction is conceived as a second-hand activity. First you do the activity - whatever it is - and then afterwards you construct a portfolio that somehow documents it. The portfolio is seen as a backward-looking reflection on the experience.

iii) A third and far richer view of the concept of the portfolio sees it as is neither a *container* nor a *reported* story, but rather as a ***dialogue***. The designer is having a conversation with him/herself (and with others) through the medium of the portfolio. So it has ideas that pop up but may appear to go nowhere – and it has good ideas that emerge from somewhere and grow into part solutions – and it has thoughts arising from others comments and reflections on the ideas. Any of these thoughts and ideas may arise from procedural prompts that are deliberately located in the activity to lubricate the dialogue. Looking in on this form of portfolio is closer to looking inside the head of the learner – revealing more of what they are thinking and feeling, and witnessing the live, real-time struggle to resolve the issues that surround and make up the task. Importantly, this dynamic version of the portfolio does not place an unreal post-activity burden on learners to reconstruct a sanitised account of the process. Creative learners are particularly resistant to what they see as such unnecessary and unconnected tasks, and this significantly accounts for their underperformance in portfolio assessments that demand such post-hoc story telling.

But real-time dynamic portfolios are not tidy, nor is it possible to present them in a pre-determined powerpoint template. It is more like a designers sketchbook – full of notes and jotting, sketches, ideas, thoughts, images, recordings and clippings. These manifestations are not random, but are tuned to the challenge of resolving the task in hand. And the point of the portfolio is that the process of working on it shapes and develops the activity and the emerging solution.

Our desire to create ***dynamic dialogue e-portfolios*** led us into all kinds of methodological problems, illustrated by the following questions.

pedagogically; what digital tools are best able to facilitate designing activity in school workshops and studios?

technologically; is it possible for learners to be developing ideas at the workshop / studio level and simultaneously reflect on the emerging web-based portfolio?
pedagogically; in development terms, is it possible for teachers to gain a sufficiently good grasp on the evolving dialogue (between studio input and web-based output) so that they can support it formatively?
functionally; in assessment terms, is it possible for teachers to gather a sufficiently rich impression of the work so as to assess it summatively and reliably?

The last of these questions – the challenge of assessment – is dealt with in detail in 3 below.

THE LEARNERS' *VOICE* IN THE PORTFOLIO

Having argued ourselves to the point of clarifying our desire for dynamic dialogue e-portfolios, we embarked on a series of activities that might best be described (methodologically) as 'play-time'. One of the problems surrounding the use of digital technologies in schools is that teachers tend towards the assumption that this needs to take place in a computer suite, rich in desktop or laptop machines where learners work with a keyboard and screen. Our starting point was very different.

We started from assumptions about the nature of design & technology – the circumstances of which are almost always workshops and studios. Two of the constants of these typical design & technology spaces are that;
– they are full of materials, apparatus, machinery, and specialist work-spaces
– they are associated with the dirty/untidy detritus of manufacturing.

They therefore make challenging locations for computers, keyboards and screens. First there is not enough space; second the space is not clean (glue, paint, flour & water, sawdust) and third learners themselves get oily or painty or gluey or floury fingers that are not then ideally suited to keyboard use. For all these reasons we took the view that peripheral, 'back-pocket' technologies would be more appropriate. At least at the 'input' level these technologies enable activities in workshops and studios to go ahead almost as normal. They don't take up too much space and (because they can be pocketed) they are not too sensitive to the clutter of the working space.

Our 'play-time' therefore involved a deliberate strategy of getting as many peripheral digital tools as we could find – taking them into design work spaces with learners – and seeing what they did with them.

One of our principal targets for this exploration was digital sound. We have long been aware that if we could capture the conversations between learners we would know far more about their own understandings about what they are doing – and why. Moreover we were aware that whilst capturing the sound digitally was quite easy, it's a very different level of challenge to display it subsequently. Does it appear in the portfolio as a sound file (to be played and listened to) or as text (converted from the digital sound). For a discussion of this specific issue see Kimbell (2005).

We took the view that speech-to-text would be a preferable approach, both for *learning* and for *assessing* reasons.

DESIGN PERFORMANCE

In learning terms, we recognise the value of learners going back over their work to modify, extend and enrich their 1st attempts. With *drawing*, we encourage them to jot down *any* graphic starting points for their designing in the knowledge that having got the process going (even with a hesitant and scrappy sketch) we can encourage them to go on and on extending and enriching their graphic explorations. In terms of learners' *writing* the same argument applies. The development and use of structuring techniques (such as writing frames) illustrates the way in which 1st ideas in writing can be re-worked to become more effective. (eg see Lewis & Wray, 1998). In exactly this way, we argued that structuring exploratory speech might also have the effect of allowing it to be refined and extended. We believed that if learners could 'see' their speech as a set of text notes on their digital design drawings – then it would help them to become more focused and purposeful with their speech.

In the confines of the TERU research space, we were able to achieve this, with learners kitted-out with headsets, and having trained the software (for at least 2 hours) to learners' individual voice profiles. But in other ways we failed with this voice-text venture:
- we could not replicate the success of the lab environment in school studios and workshops where there was much more background noise and less individual training time
- we could not achieve *conversation* between learners (even in the lab) – because the software could not readily distinguish when the 'voice' changed.

What we were trying to do was beyond (only *just* beyond) what the technology was capable of doing. In a year's time (2 at the most) increased processing power and software sophistication will make it possible. Nonetheless, we were reluctant to drop the voice from our portfolio structure and we created a pedagogic solution in association with digital voice files. Six times through the activity we asked learners to take photographs of their evolving modelling. Associated with these photo-staging-points we asked them to give us a 30 second 'sound bite' in response to 2 questions:

'what is going well with the work?'
'what needs further development?'

The 1st time they encountered this demand it typically resulted in hesitant and inarticulate responses. The 2nd time it was better and by the 6th time there were some very lucid and perceptive observations. The questions, in association with the photographs, had the effect of giving them what we might describe as a *talking* frame. As they became more familiar with it their performance rapidly improved.

At the assessment level this has proved fantastically useful, for we can now seam together the 6 voice files into a running commentary (approx 3 minutes) in which learners tell us, in their own authentic voice, the challenges they are facing and how they are overcoming them. During this 3 minutes, with the audio running commentary, the assessor can scan the other elements of the portfolio; the photographic images, the graphics and the text.

In learning terms, the value of audio-reflection has become increasingly recognised, not least through the vehicle of discussion within the learning setting.

Gardner (1983) identified the learning value of multiple ways of exploring ideas, and *discussion* highlights the role of what he characterises as *linguistic* and *interpersonal* intelligence. In the context of design & technology, we reported in the APU survey the power of group discussion as part of the activities that we created. One of the activity administrators (who managed the discussion session) commented.

> This was a strategy that I had previously not put any emphasis on in my own teaching and I found it by far the most useful device for helping pupils extend their ideas. The pupils' response to each other's criticism was a major force in shaping the success or failure of the artefact in their own eyes. Pupils saw this as a very rewarding activity and would frequently change the direction of their own thinking as a result. (Kimbell et al. 1991 p. 124)

Within the e-scape project we asked all learners to complete a questionnaire that enabled us to analyse their reaction to elements of the activity, and we specifically asked about the voice memos. In response to the statement *'The voice memos were good for explaining my ideas'* we invited them to agree/disagree on a 4 point Likert scale. Of the 256 responses (113 F: 143 M) the breakdown was as follows:

strongly agree   24 F : 50 M
agree   45 F : 63 M
disagree   30 F : 30 M
strongly disagree   14 F : 1 M

Interestingly therefore, whilst just over 60% of the females agree or strongly agree with the statement, nearly 80% of the males do. And whilst 12% of females strongly disagree, less than 1% of males do. We have yet to analyse the extent to which performance levels in the activity are affected by the voice memos, but these data suggest that we should be looking very closely at possible gender effects.

MAKING ASSESSMENT JUDGEMENTS

Of all the innovations within the e-scape project, perhaps the most far-reaching concerns the approach we have adopted to the challenge of making assessment judgements about the quality of learners' un-picked e-portfolios. We are the inheritors of a tradition in this process that broadly demands that when such assessments have to be made, we should start with assessment criteria – then allocate a set of marks to each criterion – then make a judgement about the quality of the work in relation to each criterion – and then add up the marks. What results from this process is a score for each learner that can be set against other learners' scores and we can then award grades on a scale (eg) of A-E. It is not a well-known fact that the process originated in 1792 with a Cambridge professor of chemistry – William Farish, who first attached numbers to the various parts of his student's performance and aggregated them to give an overall mark.

I have argued on many occasions (see in particular Kimbell 1997 chs 5 & 6) that this process is fraught with difficulties. Two problems in particular are paramount.

First, I do not believe that the best way to make judgements about the quality of learners' performance is by looking at bits and pieces and then adding them up. Rather I have argued that we should judge the whole piece – as a whole – and only *thereafter* drill down into that overview judgement to see how it is made up of various qualities.

Second, the approach developed by Farish assumes that if we are allocating marks - eg on a 5 point scale – for a particular quality, then there is some meaning to what a 2 might be as different from a 3 or a 4. Sometimes this supposed meaning is cast in the form of *'better'* and sometimes in the form of *'more'* and sometimes in the form of *'different'*, but in any event there is supposed to be some objective reality behind the scale.

But experts in the psychology of human judgement deny the possibility of this. In a terrific book *'Human Judgement: the eye of the beholder'* Laming (204) points out that there is no absolute judgment. "All judgments are comparisons of one thing with another" (Laming, 2004). In other words, all judgements are relative. We have only to reflect for a moment on our conventional assessment systems to see the truth of this. QCA and Examination Bodies – desperate to ensure the consistency of markers' judgements – publish extensive **exemplification** of what the marks "mean".

– this piece of work is a 3 … for these reasons (referring to the criteria)
– this piece of work is a 4 … for these reasons (referring to the criteria)

One of the problems with this position is that there will be lots of different ways of being a 3 or a 4 – depending on the balance of work across the range of criteria. But even if we forget that problem it remains the case that when markers make their judgement, they are NOT making it against an absolute objective scale, they are merely comparing the piece of work that they are trying to mark with other remembered pieces or with the published exemplars. In either event, the judgements are *comparative*.

Part of the difficulty for us in admitting this is that we have been brought up (over the last 20 years or so) to believe that assessment is *criterion referenced* and NOT about judging learners against each other. The unhappy truth, however, is that all assessment judgements involve us in comparing one learner's work with another learner's work (albeit maybe in the form of an exemplar).

If we admit the comparative nature of assessment and that these comparisons are used to derive a scale of performance (eg from A-E) it is not obviously necessary that the work should be marked in order to achieve this. The requirement is to find some way to **judge** the learners' performances in order to create the scale that is needed, and marking items to add up their scores is just the way we seem to have chosen to do this. But …

> if all judgments are comparisons of one thing with another, why do we not compare performances **directly**? (Pollitt, 2004, p 6)

This is the basis on which all (non-statistical) studies of examination comparability in England & Wales have been conducted. Most of these have been comparability studies; experiments designed to explore the equivalence of similar examinations across different Boards and/or agencies. The approach involves examiners looking at pairs of 'scripts' – each pair containing one from exam A and another from exam B - and simply reporting which of the two pieces of work is the 'better'.

The approach draws from a theory initially propounded by Thurlstone (1927), and specifically his *Law of Comparative Judgement*.

> The essential point will be familiar to anyone grounded in the principles of Rasch models: when a judge compares two performances (using their own personal 'standard' or internalised criteria) **the judge's standard cancels out**. In theory the same relative judgement is expected from any well-behaved judge. A similar effect occurs in sport: when two contestants or teams meet the 'better' team is likely to win, whatever the absolute standard of the competition and irrespective of the expectations of any judge who might be involved. (Pollitt, 2004, p 6)

Pollitt (from Thurlstone) goes on to argue that the greater the true difference between the quality of the two performances the more likely it is that the better one will win each time they are compared. Thus, a large set of comparisons does more than just generate a rank order (like FA league tables), but moreover the relative frequency of success of one performance against another also indicates how far apart they are in quality.

MODELLING THE THEORY IN PRACTICE

When we were researching the basis on which we might undertake the assessment of the e-scape portfolios, we came across this theory and were immediately attracted to it. Part of the attraction lay in the idea that the direct comparisons prioritise *holistic* judgement, but additionally it seemed that we might be able to make the whole assessment process so much simpler than is the case with prevailing practice. If all we need to do is look at a series of pairs or portfolios and decide which one in each pair (in overall terms) contains the better demonstration of capability – then that has surely got to be easier than straining over the details of deciding how many marks to award to a script for all the micro-elements of it and then adding them up.

But we were reluctant just to follow this course without first undertaking a pilot study to check the viability of the methodology. To do this I identified 20 portfolios from the previous project (*assessing design innovation*). In that project we had used a conventional assessment process of allocating marks and had ended up with a overall scale of performance (see Kimbell et al 2004). The 20 portfolios chosen from the ADI archive were selected to cover the whole of that scale and we invited Pollitt to set up the Comparative Pairs test to see whether that process would render the same rank order.

We had 6 judges from the research team and we each made 40 comparisons reporting (in each case) just which of the two pieces better met the assessment criteria we had used to identify capability. A total of 240 comparative judgements were therefore made, and the whole exercise took about 2 hours. The results of this judging were then processed by Pollitt, using a Rasch analysis. The scale that resulted ran from a low of −5 to a high of +3, the average of the 20 scripts' parameters being 0.00. The relationship between this emerging scale and the scale originally created through the numbers-based assessment process that we had used for ADI, is shown in the following chart.

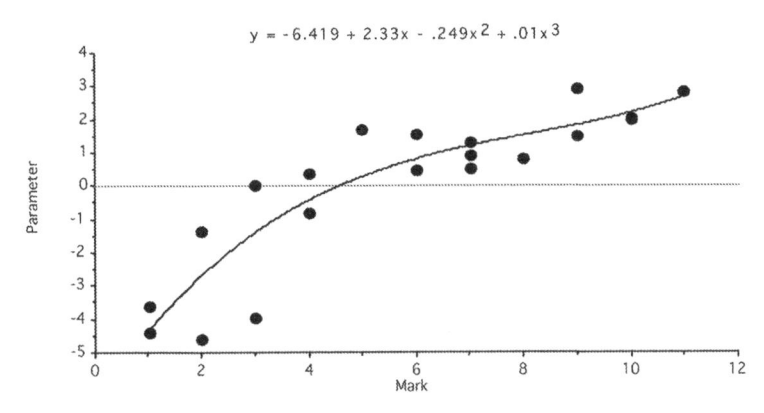

$$y = -6.419 + 2.33x - .249x^2 + .01x^3$$

Figure 1 Chart of relationship between assessment types

As expected, there is a strong but non-linear relationship between the parameters and the marks. (The relationship is expected to be non-linear because the mark scale is bounded, with a minimum of 0 and a maximum of 12, while the parameter scale runs from -∞ to +∞.) The value of R^2 was 0.81, corresponding to a correlation of 0.90 between two linear variables, as high as could be expected in a case like this. The scale reliability was estimated to be 0.92, at least as high as would be expected in a GCSE marking study. (Pollitt in Kimbell, Wheeler & Nast, 2006, p 8)

VALIDITY, RELIABILITY AND MANAGEABILITY

Any assessment system worthy of the name must match up to the demands of these three qualities.
- The system must render a **valid** measure of the capability of the candidates in the sense that experts in the field agree that the resulting scale is a 'true' reflection of the distribution of capability (the ones at the top really are better than the ones lower down).
- The system must do this **reliably** in the sense that any suitably qualified assessor will agree with the judgements made. Another word for it might be **repeatability**, so that the assessor next in line makes the same decisions that I do.

– The system must be **manageable** in the sense that it does not take too long, become too onerous, or demand any special facilities that render it impractical under 'normal' circumstances.

So, in terms of the e-scape judging, how does this Thurlstone / Pollitt comparative pairs approach measure up?

The first thing to admit is that the evidence is not yet complete. We conducted the pilot study, which gives us a great deal to work from, but (as I write) we have only just embarked on the full judging process. We have 250 portfolios in the e-scape website; we have 7 judges trained to recognise the criteria of capability; each of the judges has been supplied (from Pollitt) with a list of pairs that they have to judge; and the process began last week. Within the next 3 weeks we will have all the results back and then we can begin the analysis process. We are due to report to DfES / QCA and the Awarding Bodies in January 2007. But what – at this moment – can be said about validity, reliability and manageability?

Validity

We can examine this from the point of view of principle and of practice. In principle the process of direct, holistic, comparative judgement is a valid one. It takes the concept of holistic judgement seriously and forces the judge to balance the various qualities that might (or might not) be evident in the work. The resulting decision is not the product of some universal algorithm that values one quality above another – but is rather the product of a whole balanced judgement.

> ... why should we expect any weighted summation of micro-judgements to lead to the 'correct' macro-judgement? Given the well known complexities of weighting – the subtleties of intended and achieved weights in a composite score – it seems most unlikely that, just by chance, a total test score should happen to give the optimal measure of a student's performance or ability (Pollitt 2004 p 5).

> In principle ... direct judgement *should* be more valid than indirect scoring, since the construct that we are trying to assess becomes the actual criterion, or criteria, that examiners will use for their judgements (op cit p 14).

In practice, most measures of validity in the end come down to the judgement of expert panels. Does the panel judge that the portfolios that end up at the top of the scale really are better manifestations of design & technology capability than those at the bottom? In the pilot study – based on 20 portfolios - this was certainly the case. It remains to be seen whether this position holds true when we have completed the judging of the full e-scape sample of 250 portfolios.

Reliability

With conventional testing, reliability is a function of the number of questions or of test length (see eg. Lord, 1959), but in the *performance* domain, reliability has to be established through the assessment procedures after the test / project has been completed. With Thurlstone-style scaling, reliability is largely determined by the number of comparisons made and the number of judges involved.

In the pilot study we conducted, we had 6 judges, making 240 judgements about the 20 portfolios. In other words each portfolio was directly compared against 12 others and these judgements were distributed across a wealth of judging expertise. Compare this to a normal (eg GCSE) assessment, where a single teacher does all the marking and sometimes, this is checked by a single moderator. The reliability of Thurlstone/Pollitt judging *should* be far higher than conventional indirect scoring. Again, we shall see how the reliability of the e-scape judging emerges when we analyse the full sample.

Manageability

Why, we might ask, if this system is so simple and so clever, do we not use it already? In fact it *is* almost always used in *research studies* designed to compare one examination with another. But when we move outside the small-scale, carefully controlled research environment, it is not used at all. Why not?

The reason is simple. It demands that all the pieces of work must be available for comparison with all the other pieces of work from that examination. Imagine the logistic nightmare that that would involve. The only possibility would be to have a central hall in which all the judges and portfolios come together. Then the judges are given their lists of pairs and they fight each other for access to the ones they need. If there are thousands (or even hundreds) of portfolios, the system just becomes unmanageable.

But the rules of the game change when portfolios are based in a web-space. There, each portfolio can be accessed simultaneously by hundreds of judges – from anywhere. Provided they have internet access (broadband) and a big enough screen to scan the portfolio properly (the *e-scape* portfolios are customised to 17 inch screens) then the judging is VERY manageable. It can be done as and when is convenient in the comfort of home. Moreover, the system can have built-in, on-line support. We are currently underway with the full e-scape judging – and so far it seems to be progressing smoothly. We shall see if that remains to be true.

IN CONCLUSION

The e-scape project is unusual for us in the sense that it is entirely methodological. We are not creating all these e-portfolios and doing all this judging because we want to find out about the performance levels of learners in the sample. That was the purpose of the APU survey that I discussed at the start of this chapter, but e-scape has a different purpose. We are simply trying to develop two new methodologies:
– a new way of creating un-pickled portfolios (in a web-space)
– a new way of making assessments (Thurlstone/Pollitt judging)

The two are of course connected for without the 1st, the 2nd would be simply unmanageable. These twin challenges have resulted in any number of practical problems along the way (eg about band-width and school networks), but the theoretical problems are more interesting. I have discussed three of these theoretical problems in this chapter;
- what is an e-portfolio?
- what do we mean by the 'voice' of the learner?
- what do we mean by making assessment judgements?

In each case we have had to evolve an answer that is sufficiently convincing for the research team to press on to the next step. But we are very aware of the fact that we are still in largely uncharted territory – and there may be many dragons. We believe in our concept of dynamic conversational e-portfolios. And while we have had to compromise with the technology, we have still made our best shot at capturing the authentic voice of the learner. But it remains to be seen whether the radical new methodology for assessment renders results that are meaningful.

But the fun of research lies in thinking out ways of dealing with the inevitable uncertainty. And we always travel in hope.

REFERENCES

DfEE/QCA. (1999). *Design and technology: National curriculum for England*. In DfES (Ed.) (pp. 50): DfES Publications.

Department for Education & Skills (DFES). (2003) *Survey of information and communications technology in schools*. Oct 2003. HMSO – available at http://www.dfes.gov.uk/rsgateway/DB/SBU/b000421/index.shtml

Department for Education & Skills (DFES). 2005 *Harnessing technology: Transforming learning and children's services: DFES e-strategy* – available at www.dfes.gov.uk/publications/e-strategy

Gardner, H. (1983). *Frames of mind: The theory of multiple intelligences*. London: Heinemann.

Kimbell, R. (2005). Digital capture and the Club Med test, in *Footprints in shifting sands: Ten years of editorials from the DATA journal (1996-2005)*. D&T Association: Wellesbourne.

Kimbell, R. (1997). *Assessing technology: International trends in curriculum and assessment*. Open University Press:Buckingham.

Kimbell, R. Wheeler, T. Nast, C. (2006) *E-scape: An interim report. May 16th 2006*. TERU for QCA/DfES.

Kimbell, R., & Stables, K. (1999). *South Africa: North West Province Technology Education Project – An evaluation*. London: TERU, Goldsmiths College University of London.

Kimbell, R., & Stables, K. (2006). *Researching design learning*: Springer.

Kimbell, R., Miller, S., Bain, J., Wright, R., Wheeler, T., & Stables, K. (2004). *Assessing design innovation: A research and development project for the Department for Education & Skills (DfES) and the Qualifications and Curriculum Authority (QCA)*. London: Goldsmiths University of London.

Kimbell, R., Mahony, P., Miller, S., & Saxton, J. (1997). *Decisions by design: A research project commissioned by the Design Council* (March 1995-April 1997) Final Report. London: Goldsmiths College/Roehampton Institute.

Kimbell, R., Stables, K., Wheeler, T., Wozniak, A., & Kelly, A. V. (1991). *The assessment of performance in design and technology*. London: SEAC / HMSO.

Laming, D. (2004) *Human judgement: The eye of the beholder*. London, Thomson.

Lewis, M. and Wray, D. (1998). *Writing across the curriculum: Frames to support learning.* University of Reading: Reading and Language Information Centre.

Lord, FM.(1959) Tests of the same length do have the same standard error of measurement. *Educational and Psychological Measurement, 19,* 233-39.

Office for Standards in Education (OFSTED). (2004). *ICT in schools: The impact of government initiatives 5 years on* May 2004.. OFSTED publications – available at
http://www.ofsted.gov.uk/publications/index.cfm?fuseaction=pubs.summary&id=3652

Oxford, Cambridge and RSA (OCR) (2005) see
http://www.ocr.org.uk/Data/publications/specifications_syllabuses_and_tutors_handbooks/GCSE_Desig46205.pdf

Pollitt, A (2004). Let's stop marking exams. *Paper given at the IAEA Conference, Philadelphia,* September. Available at:
http://www.cambridgeassessment.org.uk/research/confproceedingsetc/IAEA2004AP

Prest, D. (2002). *An analysis of the attainment target level descriptors and associated programme of study in relation to the design and technology mission statement.* London: Department for Education and Skills, Design and Technology Strategy Group.

Prime Minister's Strategy Unit 2005 *Connecting the UK: The digital strategy.* A joint report with Department of Trade and Industry – available at
http://www.strategy.gov.uk/work_areas/digital_strategy/index.asp

Stables, K., & Kimbell, R. (2000). The unpickled portfolio: Pioneering performance assessment in design and technology. In R. Kimbell (Ed.), *Design and Technology International Millennium Conference* (pp. 195-202). Wellesbourne: DATA.

Thurstone, LL. (1927) A law of Comparative judgement. *Psychological Review, 34,* 273-286

Richard Kimbell
Technology Education Research Unit
Goldsmith's College
London University
United Kingdom

MARGARITA PAVLOVA

COMPARATIVE ANALYSIS AS A RESEARCH METHOD IN TECHNOLOGY EDUCATION

INTRODUCTION

This chapter explores comparative analysis as a research methodology and its application to the area of technology education. Theoretical debate on the nature of comparative education and the challenges of the modern world that should be addressed within comparative education, have emerged periodically over recent years. The debate has contributed to an understanding and shaping of the research goals, methodologies and methods to be used in the field. One of the latest 'revisions' occurred at the turn of the millennium when the special issue of Comparative Education (2000) and the subsequent special issue of responses (Comparative Education, 2001) addressed the ways comparative education is shaped by the emergence of the new challenges. As summarised by Broadfoot:

> these collections define the elements that are likely to characterise the next period of Comparative education scholarship – an engagement with the global currents of twenty-first century life; a rigorous blending of quantitative and qualitative methodologies in well-justified comparisons; a commitment to the quest for more general insights about how the key building blocks of education – culture, learning, power and technologies – work together in a context of constant change. (Broadfoot, 2003, 276)

One of the important criticisms that have emerged through discussions within comparative education relates to the scope of the debate. In particular, that the debate does not "challenge the accepted discourses of modern Western education in any profound way" (Broadfoot, 2003, p. 275). For some, the goal of comparative education is essentially a reformist one concerned with informing policy and practice in education. For others, it is simply to document the different educational realities in the world in a systematic fashion. For another group the goal is to use empirical studies as a means of generating more general understanding (Broadfoot, 2003). Novoa and Yariv-Mashal (2003) proposed a tentative chronology of the development of goals within the field, from knowing the 'other', understanding the 'other', constructing the 'other', to measuring the 'other'. They relate the current stage to 'a global climate of intense economic competition' and argue for a key role for education within it, with the major focus of comparative research "to create international tools and comparable indicators to measure the 'efficiency' and the 'quality' of education" (p. 425).

An analysis of comparative research in technology education (Pavlova, 2006) identified four stages in its development. As argued by Pavlova, the four stages could be characterised by the following statements: (1) 'help me to establish technology education', (2) 'let me tell you what's happening in my country', (3) 'this is an issue, let's compare' and (4) 'compare to establish a better society'. The difference across the approaches is concerned with goals. The goals of the first three stages are closely related to the goals of comparative education identified by Broadfoot (2003), whereas the fourth and latest stage can be seen as a response to the critique that comparative education does not challenge prevailing educational discourses. The fourth stage opens up an opportunity to use comparative analysis in technology education to establish a better society. That is, to use the results of comparative analysis as a proactive tool in challenging contemporary educational models in a profound way. Pavlova (2006) argues that stage four is a move towards the emerging stage in the development of comparative research in technology education where comparison is made on the basis of two ideological beliefs about the purposes of general education. These beliefs can be summarised as firstly, education that is designed to broaden minds and develop all students in the creation of a better society or, secondly, education that is concerned principally with training students to live and work in a market-oriented state, to be 'productive' in seizing the opportunities of the market. Pavlova, (2006) argued that a comparative analysis, using a framework based on these two ideological positions is an effective way of gaining an understanding of technology education in any particular setting. The first position can be described as developmental and the second as instrumental.

Knowing whether the rationale for technology education is based on an instrumental or developmental ideology can provide a clear understanding of the characteristics of technology education in a particular country or region. This framework for comparative research can be used to gain a deeper understanding of technology education at both theoretical and practical levels, and to identify universal and contextualised elements in approaching the area. The importance of this new framework has been justified through discussion of two research issues – values in technology education and education for sustainable development (Pavlova, 2006).

The particular goals within the field of comparative education are closely related to the methodological approaches to research. Mapping of paradigms and theories used in comparative and international education undertaken by Paulston and Liebman (1994) demonstrate a variety of approaches toward research that vary from radical-humanist to radical-functionalist, from idealist-subjective orientation or to more realist-objectivist orientation. More recent analysis by Sweeting (2005) identified the following prevailing theoretical perspectives in comparative education: Marxism/Critical Theory; Dependency Theory/World systems Analysis; Post-structuralism; Post-modernism; Post-colonialism; Feminism; Neo-liberalism/Neo-Managerialism. Ninnes and Burnett (2003) examined the historical development of comparative methodological approaches from the 1990's. They found that over the period examined, approaches moved from functionalist and positivist, towards critical, then interpretative and ethnomethodological and then finally, to post-modern. Thus, these studies demonstrate that the whole spectrum of theories have been used by researchers in comparative education. Although there is considerable variability in the way researchers position themselves in the field of comparative education, there are a number of general issues that should be considered when comparative analysis is applied.

Comparison should reflect: (a) *what is being compared with what* (e.g. teachers, schools, teaching methods and educational systems in different cultural, national and regional contexts); (b) *the evaluative basis of comparison* (e.g. the norms and principles being invoked in making comparisons); (c) *the reasons and motives underlying the comparison being made* (e.g. disinterested scholarly enquiry, a search for insights, etc., to be applied from one context to another); (d) *the methods used in making comparisons* (e.g. methods based on natural science, hermeneutic traditions, etc.) (McLaughlin, 2004).

All these general issues and concerns are applied differently in different research. This chapter analyses the methodology of one particular study to demonstrate how knowledge production in Technology Education can benefit from the use of comparative analysis.

NATURE OF THE STUDY

The study (Pavlova, 2001) was aimed at investigating understanding of knowledge in the area of technology education across four countries - Australia, Russia, the UK and the USA. Two broad questions were being asked in the inquiry: (a) how knowledge in technology education is interpreted in theory and practice in the selected countries; and (b) what conceptual framework would be suitable for future theorizing of knowledge in technology education.

Answering the first question of the inquiry: *how knowledge in technology education is interpreted in theory and practice in the selected countries,* the distinction was made between knowledge *about* technology and knowledge *within* technology. For this the nature of technology as a phenomenon was studied at the philosophical and sociological levels and then an analysis was made of how technological knowledge was presented in curriculum documents (knowledge *about* technology) and understood by key people in technology education in each country. Among the reasons for studying philosophical interpretations of technology was the lack of attention to the philosophical aspects of the phenomenon identified by several researchers (McCrory, 1987; Foster, 1992). Then, the ways of conceptualising knowledge *within* technology and technology education was also examined at the theoretical level and through an analysis of curriculum documents and the understanding of technology education expressed by experts in the field.

A comparative approach was chosen for the analysis of the concept of knowledge in technology education within the different settings, with the assumption that reflecting on the four countries would provide a broader understanding of the issues, which could lead to further development of theory in this area. The dilemmas of comparative inquiry arise from the plurality and variety of ways in which people from different countries interpret the social reality to which they belong. On the one hand, "recent cross-national research suggests that national education systems, at least as seen through official policy documents, were designed to be intelligible internationally ... [and] are easily understood by members of educational communities in other countries" (Kamens, Meyer, & Benavot, 1996), p. 122). On the other hand, similar words can have different meanings in different countries.

The investigation was aimed at examining how the phenomenon of technology is constructed differently in each of these countries. The purpose was to allow technology to be seen more clearly and to identify the particular emphases and omissions in each country, as well as identify key features of how technology education is being developed overall.

Answering the second question of the inquiry: *what conceptual framework would be suitable for future theorizing of knowledge in technology education*, this variety of ways of conceptualising and implementing technology education as a school subject and people's understanding of knowledge in it were used as a basis for further theorising. The resulting theory was presented as a model which was the major outcome of the study.

The focus for the inquiry was secondary education and it was limited to the period of one decade (1990-2000), the period when technology education was developed as a subject/learning area in general education. This study is an example of a stage three study (Pavlova, 2006), where a technology education researcher chooses a particular issue and compares it across a number of countries. This stage demonstrates an attempt to use a systematic approach towards comparison and through focusing on a particular issue, to find a better solution that is appropriate for the specific context.

METHODOLOGICAL FRAMEWORK

The methodology for this study was developed from certain initial ideas about what type of inquiry would help to understand the concept of knowledge in technology education within a comparative perspective and deal with future theorizing of knowledge in this learning area. This approach was partly the result of coming to the conclusion that there were a number of shortcomings in research in the area of technology education, where analysis was not often conceptualised within a broad theoretical framework. Dealing with the first question of this research, the concept of knowledge in technology education was analysed in relation to (1) theorizing about the nature of technology, (2) understanding the development of technology education as a field, and (3) the particular context of government documents and the understandings of key people in technology education. Implicit and explicit statements on knowledge were analysed.

As the study was located within the field of comparative education, four issues important for the methodology of this study, from a comparative education point of view, are discussed below.

(a) This study was responding to the substantial critique of the slow exploration by comparative education of the potential of post-modernity (Welch, 1997; Cowen,

1996). Comparative education as a field of study has been identified as a product of modernity (Cowen, 1996) and as a result was seen as not able to respond creatively to the "diverse and swiftly changing context" (Welch, 1997, p. 189). "The post-modern promises to situate difference centrally, including cultural diversity, could be seen as providing substantial support for a renewed form of comparative research and teaching, rendering redundant the quest for the holy grail of scientific method" (Welch, 1997, p. 187). Although, post-modern methodology was not chosen as the way of addressing the issues in this study, differences, diversity, and contexts in which a phenomenon develops were considered important. Also, a part of the analysis undertaken in this study employed modernist/post-modernist perspectives to understand the differences between countries in constructing technology education.

(b) In addressing the issue of searching for new methods of inquiry, this study was designed within the framework of qualitative research that, according to Mitter (1997), has become the growing line in comparative education methodology. Qualitative research can help to find "a fruitful balance between the messages of world system theory and the theories which regard cultural diversity to be a permanent formation of human history" (Mitter, 1997, p. 410). Comparative education has often been criticised for being 'too abstract' and for lacking empirical evidence as well as applicability in educational practice (Mitter, 1997). To overcome this 'abstractness,' interviews with key educators were considered an important part of the empirical evidence.

(c) Another issue in comparative education methodology that was relevant to this study was the uncertain nature of the unit of analysis (Cowen, 1996; Mitter, 1997). Traditional cross-national comparisons have been criticized as the role of the nation-state declines in the modern era and in the context of globalisation. As argued by Welch:

> comparative research has often failed to come to grips with the changing role of the state, and its impact on educational change... the problematising of the state, and of its role in the structuring and maintenance of power in society and education has either not been done, or not been done well in much comparative research (Welch, 1993, 12-13)

However, a number of authors (eg. Mitter, 1997) have argued that nations still play an important role in the area of education. The role of the state is increasing in aspects of social and economic policy, whereas in the nineteenth century such areas of activity were often largely directed by private, or philanthropic organization (summarized by Welch, 1993). Mitter (1997) also argued that although a 'great period' in comparative education involving the investigation of national education systems has passed:

> that does not mean, however, that this research interest will disappear in the foreseeable future, due to the continuing existence of the nation state. In spite of increasing responsibilities given to regional and local agencies and the individual schools, national curricula and school culture will go on exercising

remarkable influence on learning and socialisation processes. (Mitter, 1997, 409)

The findings of cross-cultural value surveys carried out in 44 societies at different time points, also demonstrate 'coherent and stable cross-cultural differences' across clearly distinguishable zones (Inglehart, 1997).

Multilevel analyses have been proposed by a number of researchers (e.g. Ball, 1997; Ginsburg, 1997; Mitter, 1997) as a solution to the problem. Different levels provide different perspectives on the phenomenon being examined and this increases the validity of the research. Thus, this study was located on several levels of analysis: international context of influence, national and personal perspectives. Special attention was given to considering the relationship between the universal and local dimensions, as these issues have become more central in recent studies in comparative education (Mitter, 1997).

(d) The last point to be considered here is the requirement to define the boundaries for the studied phenomenon within a comparative perspective. This raised the issue of examining technology-tekhnik concepts. The concept of tekhnik has been developed within a continental tradition: *technik* (German), *teknik* (Sweden), *tekhnika* (Russian) and are widely used in those countries. In various Encyclopedias more attention (and space) is given to defining this concept than to the concept of technology. Even though *tekhnik* and technology (in the English sense) are very close terms, they are not identical (Ströker, 1982/1989). As argued by Fores and Rey (1986), they "differ subtly but substantially" (p. 37). Terminology used in describing the phenomenon in different countries underlines the differences in the ideas behind them. So, it was important to clarify them to sharpen the focus of the study.

Historically, in continental Europe (and in Russia as well) *Technik* (concerned with making things and also an area of study in technical universities) alongside with *Wissenschaft* (literally mean 'knowledgeship', concerned with all knowledge and all subjects taught in the classical University), and *Kunst* ('art' - in the sense of 'fine arts' and the performing arts) had constituted one of the main division of academic knowledge (Fores & Rey, 1986). In contrast to that, in the UK, traditionally, science and humanities were the main division of academic knowledge. Thus, the status of technical knowledge in the Anglo-American tradition was very low. As argued by Young (1991):

> To a much greater extent than in other European cultures, technology [in English culture], and even to some extent science, has been excluded from wider discussion about culture and the school curriculum. This may reflect the perpetuation of landed aristocratic values that tend to denigrate trade and manufacture and thereby anything associated with practical skills. (Young, 1991, 235)

On the other hand, in Russia, technology (tekhnika) has been considered as one form of culture. '*Nauka*', '*tekhnika*' and '*iskusstvo*' *(science, technik and art)* were equally important. At one stage the status of technology was substantially increased and it was declared that 'technology decides/resolves everything'. As argued by Josephson (1992), in Russia they put "their faith in the so-called scientific-technological revolution and the transformation of science into a direct productive force to achieve political, economic, and social goals" (Josephson, 1992, 27).

As a result of this different understanding, school courses in technology education within the English-speaking world have been struggling for status, in particular, as a course for all. In the non-English-speaking world school courses have been constructed on the basis of a different set of assumptions. In this study three English-speaking countries and one non-English-speaking country were analysed, so this difference in approaches was taken into consideration when interpreting data.

The differences across the English and non-English speaking world in terms of *Technik* go beyond its position within academia and includes the terminology associated with it. In English-speaking countries there is a distinction between technology – technique; in Russia, between technology- technique – *tekhnika*. *Technik* has to do with the functioning of man-made things and the methods used in their manufacture (Fores & Rey, 1986). It includes "the set of particular principles according to which artifacts (human-made objects) work and the particular principles of the methods used in manufacturing them" (Fores & Rey, 1986, p. 37). This concept is not the same as the English technique. "In English the idea of technique is normally contrasted with related scientific principles, whereas *Technik* includes those principles" (Fores & Rey, 1986, p. 37). In English-speaking countries in the 19th century 'technology' meant a systematic knowledge of industrial arts with 'technique' being the means of practical application (Mitcham, 1978). In the 20th century this distinction has broken down. "It appreciates neither the inherently practical character of 'technology' (as knowledge), nor the generality of "technique" (as skill, which can be of playing the piano or even reading a book)" (Mitcham, 1978, p. 251).

In Russian and German, traditional distinctions between the terms 'Tekhnologiya/Technologie' versus 'Tekhnika/Technik' contrast processes (technology) to object (technik). When technology education was introduced into the Russian curriculum, this caused problems because technology was associated with a process. Playing chess became part of technology courses in some schools, as it is a process. The word technique is also used in the Russian language with equivalent meaning to the English word (skill in…).

Analysis of terminology is a very important part of every comparative study. It helps to highlight differences in the meanings of the same or similar concepts. For the purpose of this study 'technology' was used in the English sense as it was presented through the literature review for this study, to provide a common ground for the analysis.

These four considerations presented a framework for the study. For this study the author did not chose a particular theoretical approach although it could be considered to be close to critical theories. The approach in this study was descriptive/critical/analytical.

DATA COLLECTION

There were three major sources of data for the inquiry: a theoretical analysis of technology as a phenomenon; data from analysis of curriculum documents across four countries; and interviews with educators. The use of the multiply data-gathering strategy reflects the qualitative nature of this research. Different sources of data allow triangulation of findings across sources and strengthens the credibility of the inquiry. Triangulation of data provided a high level of authenticity to the study (Janesick, 1998). To be able to conduct the research and analysis at multiple levels, special consideration was given to the methodology of text analysis and interviews.

LITERATURE REVIEW – THEORETICAL ANALYSIS OF TECHNOLOGY

A literature review for a comparative study is not much different to other research. It also sits within the international debate. Thus, approaches to the literature review are discussed only briefly. The review of literature was formulated to review and interpret existing discussion on technology, with the aims to identify the questions which could serve as a framework for the analysis of knowledge *about* technology in the curriculum documents and interviews; to analyse theoretical understanding of knowledge *within* technology; and to spell out ideas which could be used for further theorising knowledge in technology education. Knowledge in technology education was considered as knowledge *about* technology and knowledge *within* technology. Thus, the literature review was organized in two parts. The first part, Theorizing technology, presents the discussion of ideas about technology – to demonstrate the theoretical understanding of technology as a phenomenon, and to find out the important features of technology that could constitute the basis for knowledge *about* technology (Pavlova, 2001).

The second part, Theorizing knowledge *within* technology, discussed several ways of interpreting knowledge *within* technology. Different ways of theorising technology in different disciplines (and in particular, within one discipline, the philosophy of technology) demonstrate that the emphasis placed on the nature and role of technological knowledge depends very much on what basis technology, by itself, had been analysed (Pavlova, 2005).

The literature review examined work by researchers and theorists who have presented different types of arguments, theories or questions that would seem to have relevance to the main question of the inquiry on the appropriate form of knowledge in technology education. The purpose was not to discuss how well their arguments are sustained as an overall theory, nor to consider and answer the debates about their work or between them. The purpose was to show from their writing what types of issues were raised that were relevant for the inquiry, particularly, the nature of technology and the ways of theorising knowledge in technology education.

The literature review provided a background, and set up the discussion for the last stages of the research, where the model of knowledge in technology education was developed.

METHODOLOGY OF TEXT ANALYSIS

One of the major methodological concerns for this study as a comparative inquiry was the methodology of text analysis. For this study the curriculum documents of different countries were considered an important source of evidence as to how technology and knowledge in technology education were interpreted in different countries/cultural settings and the way they were presented for teachers. The importance of texts has been argued by different authors, with very different theoretical perspectives. For example, Parsons (1970) viewed written texts as central to the process of cultural development:

> written language is a major stabilizer of culture, because it makes possible records other than the memories of participants, and an agent of more generalized diffusion in that a cultural document may be known through channels other than the particularistic communication nexus of specific persons. (Parsons, 1970, 201).

Young (1997), in considering the relationship between text and context, saw the relative freedom of texts from contexts:

> Under political conditions which permit many voices to be heard, this contextual slippage becomes the ground upon which a new message may be written. Culturally absolute claims dissolve into their own ambiguity because they are broken out of their claim to absolute embeddedness in culture by the partial separation of text from context. However, while postmodernists most often point to the problems this causes for universalist claims, they seldom show an equal awareness of the difficulty it presents for relativist views. If no text is fully embedded in context (i.e. its culture), all texts are somewhat culturally detached. All texts present opportunities for going beyond existing cultures. (Young, 1997, 504)

The ways of exploring texts have been established within the social sciences as well as within other areas of research.

'UNOBTRUSIVE METHODS' OF RESEARCH

In the social sciences the tradition of 'unobtrusive methods' (methods which do not involve talking with people) has a long history. Like any other set of methods they have their strengths and weaknesses. Textbooks and policy documents could be a source of primary data in such research. In this way libraries may supply both the data and the means to analyse it. *If the books and journals in the library are seen as data rather than as simply authoritative references, the library takes on new meanings* (Kellehear, 1993, 67).

The advantages of unobtrusive research have been well summarized by Kellehear (1993) on the basis of analyses made by Rathje (1979) and Babbie (1989). For the purpose of this inquiry the following features of unobtrusive research were important: it enables the researcher to examine curriculum documents over a decade, instead of just one cross-sectional study; it enables the researcher to see for herself in a literal sense (as opposed to self-reporting by respondents in interviews); it is not disturbing to others; and it is more easily repeated, compared to the other methods and therefore is more reliable than many other methods.

There are many techniques for interpreting patterns in data. Much depends on the intention of the reader: "*why* you are reading the texts and what do you wish to *take* from the reading" (Kellehear, 1993, 32). The purpose of research determines what is seen as important or unimportant, relevant or irrelevant. The researcher has to find a pattern in the written word to understand and explain the phenomena under study.

Kellehear (1993) summarizes three approaches to text analysis:

> Some researchers come to these sources with pre-structured categories which they have developed and look for these in the data. Others look at what themes or categories are suggested by the data itself. Or, one might attempt to discern patterns behind the obvious patterns which are suggested by the data. (Kellehear, 1993, 32)

The three approaches noted by Kellehear are 'content analysis', 'thematic analysis' and 'semiotic analysis'. This research employed the first two approaches.

CONTENT AND THEMATIC ANALYSIS

Content analysis "entails searching through one or more communications to answer questions that an investigator brings to the search" (Thomas, 1998, 119). Content analysis was used in this study to analyse the categories of technology in the curriculum documents. The intensity and quality of references to technology as a phenomenon throughout the documents were analysed. The qualitative question: "Does this document contain the characteristic for which I am searching?" was asked (Thomas, 1998, 120). In this sense the issues that interest the researcher were explored. For content analysis the categories have to be specific and clear (Scott, 1990). For this study the categories were identified through the literature review, through the analysis of the major ways of theorizing technology and its important features.

> "The most frequent criticism of content analysis is that the fetish for frequency makes the technique atomistic. This means that it breaks data into small, decontextualised and hence meaningless fragments, and then reassembles them using the researcher's own framework (as an outsider)"

(Kellehear, 1993, 37-38). Frequency was not among the indicators used in this study. To overcome this criticism thematic analysis was also used to develop an understanding by "inductively attending to the data and its source" (Kellehear, 1993, 38).

Thematic analysis was employed to analyse the data to extract categories which might reflect the understanding of the concept of knowledge in the curriculum documents and interviews. "The researcher is interested in a topic or set of issues and then approaches an interview or document with these issues in mind" (Kellehear, 1993, 38).

> Thematic analysis ... is more subjective and interpretive. Thematic analysis which takes the data itself as the orienting stimulus for analysis attempts to overcome etic (outsider's) problems of interpretation by staying close to the emic (insider's) view of the world. Also, thematic analysis does overcome the problem of viewing all events or items as having equal value or importance. Similarly, it also does not accept that frequency is, in itself, a valid or reliable indicator of importance (Kellehear, 1993, 39)

The differences and similarities between themes emerging from the analysis were of particular interest in this study.

COMPARATIVE ANALYSIS OF TEXTS

Comparative analysis of texts has not been systematically addressed in the methodology of comparative education. However, such analysis is essential to understanding the meaning of "categories and labels used by national authorities to designate the structure and content of formal schooling" (Kamens, Meyer, & Benavot, 1996, 122). For example, in the work on general methodology (*Conducting Educational Research: A Comparative View* by Thomas, 1998) several approaches to comparing texts are presented, but there is no specific discussion of the difficulties in relation to the comparative nature of this type of research. In another book *Cross-national research methods in the social sciences* (edited by Hantrais & Mangen, 1996) several difficulties in undertaking comparative research were addressed:

> The crucial difficulties revolve around specification of policy objectives and negotiating common evaluative criteria. The objectives of ostensibly similar policies vary from country to country and, indeed, from one period of history to another. A central dilemma in cross-national policy evaluation is where the performance of policies should be matched against the national (or local) objectives that spawned them or against some standardised checklist that may ignore local imperatives and conditions. At present the issue is frequently fudged if, indeed, it is recognised at all. (Walker, 1996, 148)

The question of evaluation criteria is connected to the requirement for common definitions and compatible measurements which are often unavailable. In this study, firstly, the documents were evaluated against the developed criteria (features of technology and understanding of knowledge in technology education) that emerged from the literature analysis. Secondly, the analysis was undertaken against the objectives of technology education as a learning area within each country.

There are few research reports comparing curriculum documents, but there were some studies that use textbooks as a way of examining curriculum (Altbach, & Kelly, 1988; Altbach, Kelly, Petrie, & Weis, 1991; Apple & Christian-Smith, 1991; Mbuyi, 1987).

Some scholars use content and thematic analysis. Mbuyi (1987) compares textbooks in the African contexts. Mbuyi based his analysis on the assumption that traditionally textbooks are good indicators of 'core values' transmitted to the young generation. Textbooks reflect desirable attitudes, social mores, expectations, values, and behaviour patterns. There are limitations in using texts to analyse 'school' knowledge as such an analysis does not deal with the values that children internalize from schooling.

Other authors try to use a poststructuralist approach, arguing that any text is open to multiple readings. Apple and Christian-Smith (1991) interpret Grossberg and Nelson:

> We can claim, for instance, that the meaning of a text is not necessarily intrinsic to it. As poststructuralist theories would have it, meaning is "the product of a system of differences into which the text is articulated". Thus, there is not "one text", but many... This puts into doubt any claim that one can determine the meaning and politics of a text "by a straightforward encounter with the text itself". It also raises serious questions about whether one can fully understand the text by mechanically applying any interpretive procedure. (Apple & Christian-Smith, 1991, 13-14)

Apple and Christian-Smith argue that the meanings of each text are multiple and may be contradictory and the researcher must always be willing to 'read' her/his own readings of a text and to interpret her/his own interpretations.

NON-COMPARATIVE PERSPECTIVE

Analysis of curriculum documents in this study

Curriculum documents are produced for a purpose, so it was not expected that they would include a developed and explicit theoretical presentation of underpinning assumptions. One of the main emphases in the analysis of policy texts was 'solutions to the problems' posed in each country. It was accepted that if something is explicitly mentioned it is considered a key aspect, or the writers presume that this would be a key issue for the readers, and if something was not mentioned this might be because the writers had not considered it, or it was assumed that readers and writers took this for granted. In the latter case, it was important that for any reason the writers did not feel the need to spell it out.

One methodological difficulty in comparing texts was connected to the possible contradiction between the need for consistency and the existence of other policies and texts that are in circulation (which could influence or contradict the text being analysed). 'Technology documents' constitutes a part of the broader curriculum documents and because of this, some issues common to all subjects were not discussed in the 'technology' part.

In this study only the same type of documents were discussed - those directly connected to policy in technology education. Documents for analysis had been chosen on the following basis:
- they had been published at the national level (results of national projects);
- they represented the main cornerstones in developing rationales for technology education in each country for the decade of the 1990s;
- they were published before April 1999.

The status of selected documents was different in each country. In the UK a *Statutory Order* described the frameworks for organizing technology education in state schools. In Russia, the *Standards* was not law, but it was recommended for implementation by the Ministry of Education (this means that all schools have to implement it). In Australia and the USA - the *Statement* and the *Standards* respectively, were consultative documents. In Australia, on the basis of the *Statement,* all States had developed their own curriculum and standards framework. In the USA, the status of the *Standards* was different. States may or may not use them as a framework for curriculum development. But one of the main features in the process of developing this document in the USA was the process of nation-wide consultations for consensus building among a very broad body of educators all over the country. Thus, it was possible to suggest that the rationale developed in the *Standards* in the USA would have an influence on developing policies all over the country.

Another difficulty was that the exact meaning of several categories was not clear from the documents (for example, what is meant by 'knowledge and understanding' in the UK *Order*). In this study the perspectives of interviewees were used to elucidate such issues.

The following guidelines were used in the analysis of curriculum documents:
- how technology is interpreted in curriculum documents and to what extent this understanding provides the knowledge base *about* technology;
- how technology education is understood and how it influences the concept of knowledge in technology education;
- what sort of ideas concerning the concept of knowledge *within* technology are presented in the documents.

Within these three major directions of analysis, the findings were organized around the following broad categories:
- to what extent is technology as a phenomenon described and considered as a starting point for curriculum development;
- what features of technology have been emphasized and in what traditions could it be interpreted?
- the role, aims and content of technology education;

- relationships between practice and theory, and knowledge from other areas in technology education;
- possible ways of acquisition of knowledge and its assessment;
- definition of knowledge;
- epistemological characteristics of knowledge.

METHODOLOGY OF INTERVIEWS

The qualitative nature of this comparative study was also reflected in choosing interviews conducted across countries for data collection. Interviews had been chosen as another source of evidence to construct triangulation of data. Policy-makers and key decision-makers were interviewed to understand how they interpret policy, what values they pose through the formulation of problems in technology education and the aspects of knowledge they emphasise. So, the general aims of the interviews in the framework of this study were: to provide evidence of what was happening in the four countries, and to hear the personal views of highly-positioned people in the area of technology education as a source of expert knowledge. Peoples' voices were considered as important in this research.

Ball (1997) argues that "in some policy research there are no people as such ... Both the people that 'do' policy and those who confront it are displaced" (p. 270). "By thinking about what sort of people and 'voices' inhabit the texts of policy analysis" (Ball, 1997, 271) the following 'categories' have been chosen: lecturers and professors involved in teacher education; government officials; leaders of professional associations of technology education teachers in the UK, USA and Australia (in Russia such an association does not exist); heads of curriculum development programs, innovative school educators: heads of school Technology Departments, and classroom teachers. In terms of methodological issues this range of professional occupations of the interviewees increases the validity of evidence presented. Their views represent understandings of the key issues in technology education from different angles.

The selection of interviewees and their number was based on the assumption that the key people who are closely involved in the area could provide relevant information about the policy and their perspectives are the source of valid, meaningful information. Key figures in each country were identified on the basis of their involvement in the field.

Due to the limited resources available to this study, the researcher did not have an opportunity to interview more than one educator from the USA. On the one hand, this limitation influenced the structure of research and the possibility to make a comparison between the four countries. This was acknowledged as a limitation of this study. On the other hand, the sole USA interviewee was an influential and knowledgeable member of the technology education profession, and as such, brought valuable information to this study.

The focus of the questions included: how technology and technology education had been understood in different countries (what were the concepts and theories underpinning the policies); what were the educators' attitudes to the policies; what were their understandings of the nature of the issues and problems in technology education; and what were their view on the concept of knowledge in technology education.

Semi-structured interviews were chosen as the most appropriate way of collecting data as, on the one hand, they have the advantage of directing the interviewee to focus on the crucial issues of the study and on the other hand, permit more flexibility for the interviewee to demonstrate his/her perception of reality. It is acknowledged that what was considered as an advantage in this type of research, could be a limitation in other types of research. The process of interviewing provided an opportunity for the interviewees to provide personal interpretations. That is, they were able to explore and make explicit their own understanding. The interviewer did not assume that interviewees shared her perspective or had the same framework of thought. This was stated to the interviewees prior to the interviews.

The interviews undertaken during the period of research took between 30 minutes and 1.5 hours each. One-to-one interviews were chosen, because of the following advantages: more confidential; easier to manage; and the analysis is more straight-forward.

Audio tape recording was chosen as the main method of recording interviews. In the cases where the interviewee rejected the notion of being tape recorded, notes were taken or written answers were provided (this happened twice). Interviews started with the introduction (a brief attempt to explain what the study was about), then the key areas of interest for the interview were presented: technology education as a subject and educational policy in that area; the concept of knowledge in technology education (or the theoretical model for technology education); technology as a phenomenon. Interviews then focussed progressively on more specific questions. While not all material gathered was presented in the study, all tapes have been preserved as raw data. In analysing and presenting material, detailed notes were taken focusing on key areas of knowledge in technology education. The interviews with participants in Russia were conducted in Russian. Responses were translated by the interviewer.

Several pilot interviews were conducted to understand how interviewees would interpret the questions. As a result, it became clear that an explanation was needed for the term 'technology as a phenomenon'. These responses were taken into account for the final interviews.

To interpret the results in a systematic way the following perspectives were used for the analysis:
– perspectives on technology;
– perspectives on technology education;
– perspectives on technological knowledge.

In some cases data was reported as direct quotations from the interview. The personal level data were aggregated to the country level.

The evidence presented from the interviews was limited due to the number of interviewees, limited time for follow-up interviews, and non-identical samples for each country. The results of the interviews were presented under the discussion of each country with comparison between countries.

In summary, the research methodology for this comparative study, which included a theoretical framework (developed through a literature review) and case studies of specific countries based on a series of interviews and documentary materials was designed on multiple levels. Through the combination of analysis of theoretical emphases in different countries with the presentation of information-rich cases from each country, an optimum balance between in-depth and broad inquiry was achieved.

Although the question of reliability of methodology and validity of interpretations was specified among the non-fully addressed issues in qualitative research in comparative studies (Walker, 1996), for this inquiry, validity was supported by explicit presentation of data from a number of sources and the reliability of the research was strengthened by careful documentation of the process of study and by the use of texts as the source of evidence. However, the impact of the researcher on the interpretations that were made and on the nature of the data collected was recognised. A descriptive and analytic presentation of data from the interviews and documents was used for the further conceptual development of the framework for the analysis of knowledge in technology education.

RESULTS

The results of this study were reported in a number of publications (see for example, Pavlova, 2002, 2002/2003, 2003, 2005), thus here only the model *Knowledge in technology education* (Figure 1) is presented. This research was generated with the aim of understanding the nature and place of technological knowledge in technology education. In the context of this study it is important to note the absence of previous research which seeks to compare the understanding of technological knowledge between several countries, including non-English speaking countries. This study contributes towards filling this void. Research demonstrates that the concept of knowledge in technology education has not been developed in theory or understood in practice in an holistic, systematic way. There is no agreed framework for theoretical analysis of technological knowledge. Many participants in this research emphasised the importance of developing this concept further.

It is possible to suggest that the absence of agreed frameworks has occurred because of two sets of problems. One of them is connected to the non-coherent way of theorising technology in non-educational contexts and the other with the under-developed theory of technology education as a learning area.

KNOWLEDGE IN TECHNOLOGY EDUCATION

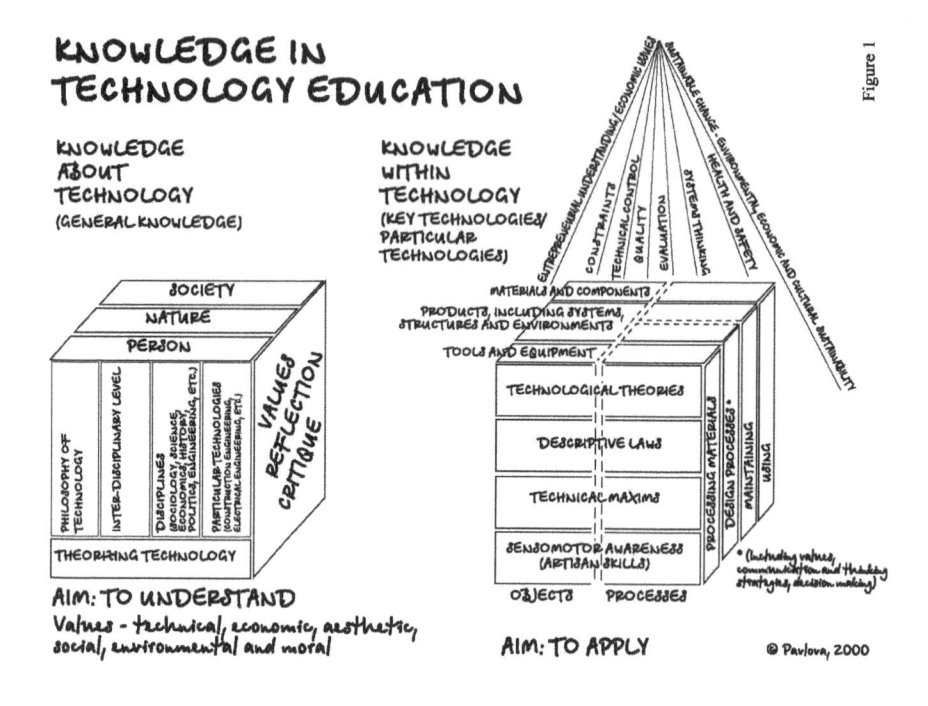

Figure 1. Pavlova, M. (2005). Knowledge and Values in Technology Education. In: International Journal of Technology and Design Education, 15(2).

This study demonstrates that knowledge *about* technology and knowledge *within* technology constitute two important elements of technological knowledge and should be reflected in curriculum.

CONCLUSIONS

As is the case for other areas, comparative research in technology education provides a unique opportunity for considering complex issues within technology education across a number of countries. The production of academic knowledge in the modern world is seen as an 'interweaving of contrary currents' – of 'internationalisation' and 'indigenisation', of global diffusion processes and culture-specific reception processes, and of the global spread of standardised models and the persistence of diverse socio-cultural configurations (Schriewer, 2004). These 'global currents' influence the ways we study education.

Examples from the study discussed in this chapter demonstrate that the methodology of comparative education can be used effectively in technology education and it provides a deeper understanding of the issue being examined, compared to the analysis that might be achieved within one setting. Such issues as goals for comparison (the challenge of accepted discourses in modern Western education), units of analysis (multiple levels of analysis), boundaries for research (cross-cultural examination of phenomena under inquiry), terminology (exploring the meanings), frameworks for analysis (a rigorous blending of quantitative and qualitative methodologies and theoretical perspectives), and methods of inquiry (texts analysis, interviews, surveys, etc) should be carefully considered when a study is designed. Application of the above considerations for the methodological design for a particular study can be helpful for someone who is interested in a broad understanding of issues in technology education and who is willing to explore different perspectives in developing a theory further.

REFERENCES

Altbach, P. G. & Kelly, G. P. (Eds.) (1988). *Textbooks in the Third World: Policy, content and context.* New York, NY: Garland Publishing.

Altbach, P. G., Kelly, G. P., Petrie H. G., & Weis, L. (Eds.) (1991). *Textbooks in American society: Politics, policy, and pedagogy.* Albany, NY: State University of New York Press.

Apple, M. W. & Christian-Smith, L. K. (Eds.). (1991). *The politics of the textbook.* New York, NY: Routledge.

Babbie, E. (1989). *The practice of social research* (5th ed.). Belmont, CA: Wadsworth.

Ball, S. J. (1997). Policy sociology and critical social research: A personal review of recent education policy and policy research. *British Educational Research Journal, 23* (3), 257-274.

Broadfoot, P. (2003). Editorial. post-comparative education? *Comparative Education, 39* (3), 275-278.

Cowen, R. (1996). Editorial. *Comparative Education, 32* (2), 149-150.

Fores, M. J. & Rey, L. (1986). Technik: The relevance of a missing concept. In A. Cross & R. McCormick (Eds.), *Technology in schools* (pp. 36-48). Milton Keynes, England: Open University Press.

Foster, W. T. (1992). Topics and methods of recent graduate research in industrial education. *Journal of Industrial Teacher Education, 30* (1), 59-72.

Hantrais, L. & Mangen, S. (Eds.) (1996). *Cross-national research methods in the social sciences.* London: PINTER.

Inglehart, R. (1997). *Modernization and postmodernization: Cultural, economic, and political change in 43 societies.* Princeton, NJ: Princeton University Press.

Janesick, V. J. (1998). *"Stretching" exercises for qualitative researchers.* Thousand Oaks, CA: Sage.

Josephson, P. R. (1992). Science and technology as panacea in Gorbachev's Russia. In J. P. Scanlan (Ed.), *Technology, culture, and development: The experience of the soviet model* (pp. 25-61). Armonk, New York: M. E. Sharpe.

Kamens, D. H., Meyer, J. W., & Benavot, A. (1996). Worldwide patterns in academic secondary education curricula. *Comparative education Review, 40* (2), 116-138.

Kellehear, A. (1993). *The unobtrusive researcher: A guide to methods.* St. Leonards, NSW, Australia: Allen & Unwin.

Mbuyi, D. M. (1987). *Beyond policy and language choice: An analysis of texts in four instructional contexts in East Africa* (Special Studies in Comparative education, 18). Buffalo, New York: Comparative education Center State University of New York at Buffalo.

McCrory, D. L. (1987). *Technology education: Industrial arts in transition, a review and synthesis of the research* (4th ed.). Columbus, OH: the National Center for Research in Vocational Education, the Ohio State University.

McLaughlin, T. H. (2004). Education, philosophy and the comparative perspective. *Comparative Education, 40* (4), 471 - 483.

Mitcham, C. (1978). Types of technology. *Research in Philosophy and Technology, 1,* 229-294.

Mitter, W. (1997). Challenges to comparative education. Between retrospect and expectation. *International Review of Education, 43* (5-6), 401-412.
Ninnes, P & Burnett, G. (2003). Comparative education research: Poststructalist possibilities. *Comparative Education, 39* (3), 279 - 297.
Nóvoa, A. & Yariv-Mashal, T. (2003) Comparative research in education: A mode of governance or a historical journey? *Comparative Education, 39* (4), 423 –438.
Parsons, T. (1970). Some considerations on the comparative sociology. In J. Fischer (Ed.), *The social sciences and the comparative study of educational systems*. (pp. 201-220). Seranton, Pennsylvania: International Textbook Company.
Pavlova, M. (2006). Comparing perspectives: Comparative research in technology education. In M. de Vries & I. Mottier (Eds.) *International handbook of technology education: The state of the art*. Rotterdam: Sense Publishers.
Pavlova, M. (2005). Knowledge and values in technology education. *International Journal of Technology and Design Education, 15*(2), 127 – 147.
Pavlova, M. (2003) Technological culture: Its representations and potential for theorising technology education, *The Canadian Journal of Science, Mathematics and Technology education, 3* (1), 37 –49.
Pavlova, M. (2002/2003) A comparison of standards for technology education in Russia and America. *The Technology Teacher e* (www.iteawww.org/mbrsonly/TTTE/F3.html)
Pavlova, M. (2002). Teaching design: aesthetic, cognitive or moral emphasis? *Design and Education, 9* (1), 5-18.
Pavlova, M. (2001). *Theorizing knowledge in technology education: Policy analysis of four countries*. Unpublished PhD Thesis, La Trobe University, Melbourne.
Paulston, R. G. & Liebman, M. (1994). An invitation to postmodern social cartography. *Comparative Education Review, 38* (2), 215 –232.
Rathje, W. L. (1979). Trace measures. In L. Sechrest (Ed.), *Unobtrusive measurement today* (pp. 75-91). San Francisco: Jossey-Bass.
Schriewer, J. (2004). Multiple internationalities: The emergence of a world-level ideology and the persistence of idiosyncratic world-views. In J. Schriewer (Ed.), *Multiple internationalities: The emergence of a world-level ideology and the persistence of idiosyncratic world-views* (pp.473 - 533). Frankfurt a. M. & New York: Campus.
Scott, J. (1990). *A matter of record*. Cambridge: Polity Press.
Ströker, E. (1989). Filosofija tehniki: Trudnosti odnoj filosofskoj distsiplinu [Philosophy of Technik: The difficulties of one philosophical discipline]. In T. G. Arzakanjan & V. G. Gorohova (Eds. and Trans.), *Filosofija tehniki v PhRG* (pp. 54-68). Moscow: Progress. (Original work published 1982)
Sweeting, A. (2005). The historical dimension: A contribution to conversation about theory and methodology in comparative education. *Comparative Education, 41* (1), 25 –44.
Thomas, R. M. (1998). *Conducting educational research: A comparative view*. Westport, CT: Bergin & Garvey.
Walker, R. (1996). Part four: Evaluation. In L. Hantrais & S. Mangen (Eds.), *Cross-national research methods in the social sciences* (pp. 148-151). London: PINTER.
Welch, A. R. (1993). Class, culture and the state in comparative education: Problems, perspectives and prospects. *Comparative Education, 29* (1), 7-28.
Young, M. (1991). Technology as an educational issue: Why it is so difficult and why it is so important. In H. Mackay, M. Young, & J. Beynon (Eds.), *Understanding technology in education* (pp. 234-243). London: The Falmer Press.
Young, R. (1997). Comparative methodology and postmodern relativism. *International Review of Education, 43* (5-6), 497–505.

Margarita Pavlova
Griffith Institute for Educational Research
Griffith University
Australia

KAY STABLES

OBSERVATIONAL TECHNIQUES FOR EXAMINING STUDENT LEARNING ACTIVITY IN TECHNOLOGY EDUCATION

INTRODUCTION

This chapter provides insights into the issues involved in creating an observation framework for exploring the detailed actions that take place when learners are engaged in technological activities. It takes as its starting point a case study of how such a framework was established for the Understanding Technological Approaches Project (1992 - 1994). This project was undertaken in England in the early years of the introduction of its National Curriculum, by a research team in the Technology Education Research Unit (TERU) at Goldsmiths College University of London. It was funded by the Economic and Social Research Council (ESRC) of the UK.

The project followed directly after the completion of TERU's major Assessment of Performance in Design and Technology (APU D&T) project (Kimbell et al., 1991). In this project we had undertaken, for the national government, a survey of the design and technological capability of a 2% sample of the 15 year olds in England, Wales and Northern Ireland – a sample of about 10,000 learners. The APU D&T project had focused on assessing outcomes derived from (predominantly) short focused activities and supplemented by a series of case studies assessing the outcomes of full-length projects. The research forming the case study in this chapter sought to fill two specific information gaps left by the APU D&T project: first, insight into learner activity and approaches throughout a design and technological project; and second, whether and how this differs for learners of different ages. The team observed in detail the complete projects of 80 learners, representing all age groups from 5 to 16, engaged in projects lasting from 3 to 48 hours.

The chapter describes the methodology for both creating and using the framework and for analysing the data derived from it. It discusses the issues involved in observing complex, integrative, authentic activity, in detail, over large expanses of time and with different age groups. It also describes how this observation methodology was developed, adapted and utilised in a number of research endeavours since the initial project, and gives brief consideration to related approaches to observation in research.

H.E. Middleton (ed.), Researching Technology Education, 135–154.
© 2008 Sense Publishers. All rights reserved.

EXPLORING AUTHENTIC ACTIVITY, OBSERVING 'REAL' PERFORMANCE

The APU D&T project had been explicitly committed to creating activities that were as 'real' as possible, both in the design and technological challenge they presented and in the way in which the activities were structured. During the project we had explicated a model of design and technological activity that denied traditional linear and cyclical models, made up of clear and prescriptive 'steps' such as identifying problems, conducting research, generating ideas and so on, that prescribed the route through an activity. Instead we proposed a model that was driven by the generation of ideas and that was progressed through an iteration of action and reflection aimed at developing ideas to the point where they became working realities. Thus steps were not prescribed in advance but rather emerged as appropriate responses to the development of an idea. We believed (and still do) that this model was more authentic as a description of the reality of dealing with the complexity and uncertainty of the 'wicked' (Buchanan, 1995) problems of designing.

While we took this model as our guiding principle in creating authentic assessment activities, we had little evidence of the way things played out in regular, full-length projects in design and technology classrooms. Within the APU D&T project we had conducted case studies of extended project work that was being undertaken for the newly developed GCSE (16+) examinations in design and technology related subjects. But while this had given us experience of using a performance framework for looking at project work over time – we had been looking at the *outcomes* of the work (drawings, models and so on) at intermittent stages through the project, not at the actual performance that had created them. Having completed the APU D&T project it was this *real-time* performance data that we sought, and – most importantly – across all ages within compulsory schooling, not just with 15 year olds.

UNDERSTANDING TECHNOLOGICAL APPROACHES

To fill this information gap, Richard Kimbell (the director of the project) secured funding to undertake a new project – the Understanding Technological Approaches project – in which our initial idea was to replicate the APU D&T case studies, but with a broader age group. Early trials showed that this approach –visiting schools on three occasions during a project, and reviewing with learners the development of their project –would be quite inadequate to portray critical features of learners' activity and approaches. Particular difficulties emerged where there was little tangible evidence (drawings, notes etc), and where the learners found it challenging to remember exactly what they had done between visits – and more importantly why. We decided we had to be present in the classroom and observe activity as it played out. This resulted in us shifting from a broadly interview-focused research strategy to a broadly observational one. Denscombe (2003), in a useful discussion of using observation in research, sets out in a straightforward way that which our research need had become.

> Observation offers the social researcher a distinctive way of collecting data. It does not rely on what people *say* they do, or what they *say* they think. It is more direct than that. Instead, it draws on the direct evidence of the eye to witness events at first hand. It is based on the premise that, for certain purposes, it is best to observe what actually happens. (Denscombe, 2003, 192)

In shifting the research approach at this early stage in the project we needed to be mindful of what we had initially told our sponsors about how we would conduct the research, and the level of resources they had provided us with to do it. Addressing the first of these was a validity issue – we believed that detailed observation would give us 'truer' and thus more valuable data. We trusted the sponsors would be supportive of us undertaking a more valid approach than that of us blindly following the plan we had made in advance – and indeed this was the case. The second issue of resources was a more pragmatic challenge – visiting a school on three occasions is far less resource hungry than being present to observe actual performance. In addressing this, we added into the equation reliability – how much, of how many projects, did we need to observe to answer our research questions?

The shift in approach also illustrates certain similarities between designing and researching. Research problems (certainly in socio-cultural settings) are 'wicked' problems and addressing them requires a responsive, iterative process, as is recognised by researchers outside of design or technology education who subscribe to the 'Design Experiments' approach to research (see Kelly, 2003).

> Just as the design activity for an artifact, intervention or initiative is a creative process so too is the ensuing research process. As a result it is difficult to provide a detailed outline of the procedures one must follow to do a design experiment. (Gorard & Taylor, 2004, 108)

CREATING THE FRAMEWORK

Having decided that we wanted to observe performance 'for real', our first task was to create a structure within which to record our observations. Our approach to this was partly driven by our concern to understand the ways in which designing is pursued (and hence derived) from the processes of designing, and partly from our interest in learning and teaching. Pragmatism was once again part of the equation – what could we actually observe and record without distorting the reality of the classroom and within our given resources. Literature on using observation in research settings such as classrooms makes a distinction between broadly two approaches – *systematic* observation and *participant* observation. Systematic observation has its roots in social psychology and tends to collect quantitative data, sees the researcher as a 'fly on the wall' and undertakes observation against a pre-defined checklist or observation schedule. Participant observation is more closely associated with sociology and anthropology and aims to get 'inside' the social and culture realities of the situation being researched by the observer literally becoming a participant in the situation. It tends to produce qualitative data.

Our own approach evolved through early explorations that resulted in us adopting certain characteristics of both types. We started with a basic set of research 'needs' – we wanted to know what the learners did, why they did it, when they did it and what interactions took place between them and their teachers. Our starting point was literally to go into the classroom and try things out, starting with various time blocks (from one minute to 5 minutes to 15 minutes) and by taking notes as a free text narrative. However, we weren't starting from scratch in terms of what we thought was important in terms of design and technological capability. For example, we knew from the APU D&T survey that the grip the learner had on the task and the way they 'grew' their ideas was important, and we believed that the extent to which they were in 'active' or 'reflective' mode would be significant.

Through experimenting with observation approaches in classrooms and through discussions within the research team (of three) we found that observing 4 learners every 5 minutes was manageable and gave a good picture of what was taking place. We also realised that we could optimise the time available by pre-coding certain elements we wished to collect data on and that it was also critical to record a free text narrative that contextualised and elaborated on the coded data.

In developing a pre-coding system we were moving towards the systematic observation paradigm – producing data that was quick to collect and that facilitated quantitative analysis. Despite having certain areas to target in the coding, creating a system we were comfortable with in terms of its authenticity for providing insights into the processes underway (designing, learning and teaching) took some time and went through several iterations. Coding systems, and the 'checklists' they produce have a tendency to appear straightforward and simple – but below the surface lies a potential minefield of problems – as we are reminded by Seale.

> Coding is ... the first step towards data analysis. Decisions taken at this stage in a research project have important consequences. The quality of a coding scheme influences the quality of data analysis, for it is in coding schemes that a researcher becomes committed to particular ways of categorizing the world. Coding schemes can be narrow, artificial devices that hinder thought, or they can contain the seeds of creative new insights. (Seale, 2004, 306)

Identifying key areas to code

At the core of our interests were the design *intentions* of the learners and developing a coding system for this was relatively straightforward, as design intentions relate directly to dimensions of design process. If your intention is to generate ideas, the processes you employ are those for generating ideas; if your intention is to evaluate your designing, then the processes you employ are evaluative. Listed in figure 1 are the intentions we coded, together with the working definitions used by the research team in our observations.

CLASSROOM OBSERVATIONS

Intentions - NWhy?Ò (at least one thing always filled in)
Identifying:	needs, issues, tasks, etc
Generating:	Any point in the task when the child comes up with a new idea.
Modelling: exploring:	When the intention is to explore an idea they have generated
Modelling: developing:	When the intention is to develop an idea they have generated
Modelling: modify:	When the intention is to modifying an idea they have already developed.
Detailing:	When the intention is to work out the detail of an idea.
Constructing:	When the intention is to make the design for real.
Planning:	When the intention is to work something out in advance of doing it.
Organising:	as in sorting out folder, work space.
Investigating:	When the intention is to find something out to assist with any aspect of the designing or making.
Seeking help:	From the teacher - clarification, direct instruction, preparation of materials etc.
Receiving:	When the intention is to receive knowledge, skill or understanding that has not been sought in advance.
Evaluating:	When the intention is to make a judgment or decision about any aspect of the design as it develops.
Reviewing:	When the intention is to check on progress in any aspect of the task.
Explaining:	Explaining their work to other pupils, teachers, researcher, visitor etc.
Presenting:	their work, either in their folders or to others.
Recording:	When the intention is to record any aspect of the designing or making.
Intentionless:	When the child is off task in terms of D&T

Figure 1 Working definitions of design intentions for coding

There are different ways in which any process (and indeed intention) can be undertaken – for example, if your intention is to generate ideas, you might do some drawings, you might engage in discussion, you might manipulate materials. Thus we developed a coding system for recording *what* learners were doing – the *manifestations* of their intentions – were they drawing, writing, talking, cutting materials, mixing materials and so on. This set of descriptors we derived in part from our cumulative experience as design and technology educators and in part from classroom observation – whenever we witnessed a learner doing something new it was added to the list.

Manifestations - NWhat?Ó (at least one thing always filled in)

Discussing:	When the child is talking about some aspect of their work with another person.
Talking:	When a child is making a statement, asking/answering a question - not discussing
Thinking aloud:	When the child is talking through their thoughts, not necessarily to another person.
Listening:	When the child is listening to instructions, advice etc (usually from the teacher).
Looking:	When the child is looking at or watching another person/product/process.
Drawing:	
Reading:	
Writing:	
Waiting:	Waiting for the teacherÕs support/guidance.
Making - cutting:	Separating materials with tools such as fingers, scissors, saws, files, knives, wire, punches,drills, spatulas
Making - joining:	Joining materials together - stitching, gluing, nailing, screwing, welding, soldering,
Making - fitting:	Shaping and/or joining materials so that they fit together or make the product fit to something else, such as a person's body, another part of a system.
Making - moulding:	Shaping material.
Making - mixing:	Mixing materials together such as foods, paints, dyes.
Making - finishing - base:	Adding finish to a surface by working on the base material eg polishing, engraving, sanding.
Making - finishing - add:	Adding a separate finish to a material eg enamelling, embroidering, printing
Arranging:	Materials.
Selecting:	Making choices eg of materials, tools, design ideas, effects.
Measuring:	Ascertaining the weight, height, length etc of something using standard or non standard units.
Marking out:	Marking critical points on a material where things will be cut or joined eg by use of templates or paper patterns, pencils, pens or scribes.
Preparing:	Getting tools, materials or working areas ready.
Testing:	
Cleaning up:	
Off tas k:	Not engaged in D&T activity.

Figure 2 Working definitions of manifestations for coding

To further qualify the 'what' and the 'why' of the learner's actions, we wished to know what the focus of their work was at any one time: were they dealing with *task* issues, such as the needs of the user; were they focusing on *communicating* their idea – to themselves or others; or were they focused on manufacture, or *manufacturing* issues? These items became the next category for coding.

Issues child is dealing with

T = Task issues:	The child is directly concerned to handle, through their designing and making, issues to do with the way the outcome interacts with the user.
C = Communication issues:	The child is directly concerned with the way their design ideas are being communicated, either to themselves, or to other parties.
M = Making issues:	The child is directly concerned with issues to do with making their design work, such as the tools, materials and techniques being used.

Figure 3 Working definitions of issues for coding

As we were observing within a learning and teaching context, it was also critical that we had some way of identifying the nature of the interactions between the teacher and the learner. As our ultimate concern as educators is developing the capability of the learner, of particular interest was learner autonomy – who was driving the action? To this end we noted any teacher-learner interactions in each 5 minutes – and where they occurred, was the teacher in *directive* or *supportive* mode.

Teacher intervention (inc. Researcher) (filled in as appropriate)
dir = direction: Teacher intervention to give direction to the child.
sup = support: Teacher intervention to support child's own ideas/work.

Figure 4 Working definitions of direction and support for coding

We believed it would be important to get some measure of the learner's engagement of the task and were quietly surprised when early trials quickly indicated how easily and reliably we could gauge this by considering the *pace* at which they were working in any 5-minute block of time. To identify levels of pace, we drew on the metaphor of driving. Observing any one learner, it was easy to see if they were fully focussed and really *motoring* along with their work, as opposed to being on task, but just *poddling* along or being completely off-task altogether – *stationary*. These three terms allowed the team to share an understanding of pace quickly – and from our observation there was a direct link between a learner's pace of work and their engagement with it.

Level of engagement/motivation/pace in task (always filled in)
S = Stationary: The child is effectively 'off task'.
P = Poddling: The child is working at their task, but not giving it full attention or all their energy.
M = Motoring: The child is fully engaged in the task and is giving it full attention, energy and commitment.
NB When teacher is talking to the class as a whole, or to an individual, fill in poddling, unless the pupil is very dynamic or clearly not listening.

Figure 5 Working definitions of pace for coding

The terms were not just drawn from the metaphor, but also from the vernacular – poddling perhaps being the most extreme, as it seems to be a word understood more readily by those from England. From the point of view of reliably collecting the data, shared understandings within the team were paramount and the use of metaphor and the vernacular greatly assisted this – a tactic we have found valuable in other research projects (Stables & Kimbell, 2006). The final version of the 5 minute recording section of the observation framework is shown in figure 6.

KAY STABLES

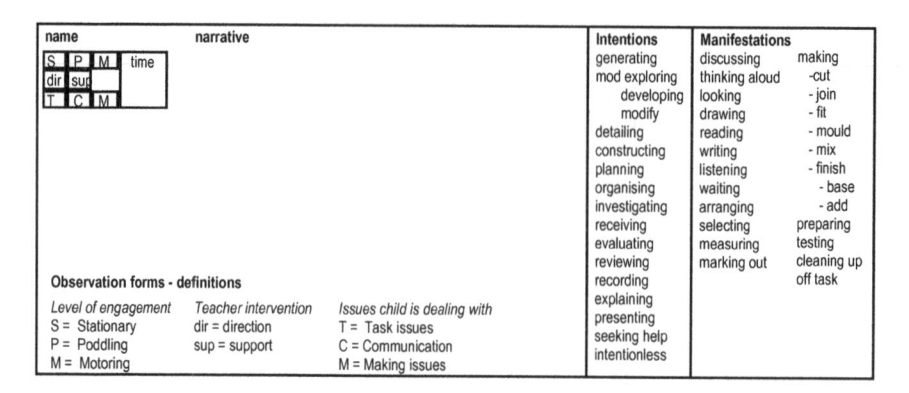

name			narrative		Intentions	Manifestations	
S P M time					generating	discussing	making
dir sup					mod exploring	thinking aloud	-cut
T C M					developing	looking	- join
					modify	drawing	- fit
					detailing	reading	- mould
					constructing	writing	- mix
					planning	listening	- finish
					organising	waiting	- base
					investigating	arranging	- add
					receiving	selecting	preparing
					evaluating	measuring	testing
					reviewing	marking out	cleaning up
Observation forms - definitions					recording		off task
Level of engagement	Teacher intervention		Issues child is dealing with		explaining		
S = Stationary	dir = direction		T = Task issues		presenting		
P = Poddling	sup = support		C = Communication		seeking help		
M = Motoring			M = Making issues		intentionless		

Figure 6 the final observation framework

The importance of narrative free text

Creating an observation framework, including the categories and pre-coding shown above, made it both easy and quick to gather a considerable amount of quantitative data in each time frame. But experience showed us that if all we collected was the pre-coded data we had no way of really knowing what was happening – we needed ongoing, narrative, qualitative data that contextualised the coding and also that allowed us to cross moderate the coding, as we did on a small sample of each researcher's work. So in addition to the coded data we had the free text that recorded a combination of a description of events and also observer annotation and thoughts. In reality, this meant that sometimes there was a considerable amount of free text – typically when observing the younger children who were *motoring* and who undertook an impressive range of actions in five minutes, while in others there was very little. At its most extreme, a Year 10 learner intent in achieving a particularly good finish on a wooden clock, provoked a narrative record stating "sanding", "still sanding", "yet more sanding" etc for close to an hour of work time.

USING THE FRAMEWORK

Having created the framework through initial trial observations, there were a number of issues that had to be addressed in terms of how we would use the framework for the main fieldwork.

Participant or non-participant observation

A critical issue in using the framework was whether we saw ourselves as non-participant observers, in line with the systematic observation 'fly-on-the-wall' approach, or as participants. We were keen to have as little influence as possible on what was happening in the classroom, so lent towards the non-participant approach. However, we were also keen to ensure that our data was as accurate as possible and, while the *manifestations, pace* and *interaction* data was relatively straightforward in this respect, the design intentions were not. We therefore decided that, at times, we would need to interact with learners to be sure we understood their intentions – it being better to ask than to assume. We adopted an explicitly neutral approach to this, whereby we would ask what the learner was doing and why, but never indicate whether we thought this a good or bad thing to be doing, or provide an alternative suggestion. In order to optimise the reliability and validity of the *data*, it could be argued that we compromised on the method. But in our view we took an intelligent approach that accepted and acknowledged the reality of the research situation. Thus we knew our presence would have an effect on the learners being observed, possibly a 'halo' effect, so we acknowledged this and did our utmost to minimise it, whilst still seeking good information. We knew, particularly with younger children, any extra adult in the classroom would be seen as fair game to provide help, so if they approached us, we interacted with them only in an entirely non-directive manner. As we reported to the sponsors at the end of the project,

> The observation procedure required the pupil to know they were being monitored, and at intervals the observer might supplement their observations by interacting with the pupils to ask about some feature of their current work. In primary schools we found this interaction to be an inevitable part of our presence in the room. The pupils had no hesitation in putting us to good use as helpers and general consultants. In secondary schools we found that it is more possible to operate as 'a fly on the wall', but nevertheless there were times when it was valuable to get comments from the pupils. Whenever these interactions took place, it was always on the basis of ***neutral*** questions by the observer. (Kimbell et al., 1994, p. 4)

In addition to the quality of the data derived, the position we adopted needed to be considered in terms of research ethics. As researchers we wanted to be as open and honest as possible, to respect the rights and dignity of the learners and teachers we were observing, and, critically, to avoid harming the participants in any way as a result of the research. While the research clearly posed no physical threat, our interactions, or a denial to interact, could have been seen as a pedagogic threat. In remaining neutral our aim was to be non-threatening to the learner, their motivation and their progress.

Reliability issues

As indicated in outlining the way the observation framework was created, the research team was small and the framework was developed through classroom trials and hours of discussion within the team, all of which contributed to the overall reliability of the data collected. But it is a fact of life that two people observing the same situation will see things differently. In discussing this issue, Denscombe points to the way individuals 'filter' observation by selectively perceiving the situation and selectively recalling the situation. Also, the filters are influenced by personal sensitivities, such as hunger, anger or frustration. He indicates three dimensions that play a key role:

> The selection and organization of stimuli ... is far from random. In fact there is a tendency to highlight some information and reject some other, depending on:
>
> *Familiarity*. We tend to see what we are *used* to seeing. If there is any ambiguity in what is being observed, we tend to interpret things according to frequent past experiences
>
> *Past experiences*. Past experience 'teaches' us to filter out certain 'nasty' stimuli (avoidance learning) or exaggerate desirable things.
>
> *Current state*. Physical and emotional states can affect what researchers perceive. Physiological states such as hunger and thirst can influence the way we interpret what we 'see'. Emotions, anxieties and current priorities can likewise alter our perceptions. (Denscombe, 2003, 194)

His point here is that, as human beings, we each interpret what we see in our own way, so our task in both creating and using the observation framework was to minimise the interpretation of individual researchers. By creating the pre-coded categorisation and items we provided clear targets for the observation, which in themselves aided reliability, as each researcher was 'looking' in the same direction. We also undertook exercises whereby we each observed the same learner at the same time, and then compared notes, identified discrepancies and ironed out differences in interpretation. It was through these exercises that we created guidance notes and working definitions to strengthen inter-observation reliability. In addition, as already mentioned, the narrative descriptions allowed for moderation. Finally, we supported our observation findings with (taped) interviews with learners at the conclusion of the project. We also gathered contextual information about the school, the department (in the case of secondary), the teacher and the learner, all collected through discussion with teachers involved.

Designing the sample

In preparing to use the framework we also had to design the sample to be observed. We had established that we could comfortably observe 4 learners at any one time, and so we needed to specify how these learners would be selected. In order to see a range of responses, we asked the teachers to choose the learners based on selecting the best designer/technologist in the group, two middling performers and one learner of low general ability but who was good at design and technology. We also asked the teachers to balance the group in terms of gender and, for pragmatic reasons, to only select learners who were good attenders.

We wished to gather data across the full age range of compulsory schooling (5-16). We suspected that there was likely to be particularly interesting data derived from primary / secondary transition and so clustered a number of observations in Years 6 (primary), 7 & 8 (secondary). Because the National Curriculum specified areas of design and technology (resistant materials, systems and control, food and textiles) we identified projects that would cumulatively, allow us to see the full range of materials in use. We were also keen to research in a range of school types, and from different areas. All of the above considerations were targeted at the data being derived from as full as possible a range of design and technology settings. Our final decisions were more directly to do with manageability – maximising the resource we had to the number of schools we could include. This resulted in us choosing schools within a 60-mile radius of London (our base), and nearer for schools where the projects were long and would require a considerable number of visits.

The final sample involved 80 learners across 20 schools and projects ranging from 3 hours (with Year 1 learners) to 48 hours (with Year 11), each of which we observed, 5 minutes by 5 minutes, for the full project length.

ANALYSING DATA DERIVED

By the end of the fieldwork we had a vast amount of data on each individual learner and our first task was to organise the data in a way that enabled us to analyse it. Our starting point was simply to enter the data into a Microsoft Excel spreadsheet, taking a single line for each learner. Once entered in this way we could review the 'journey' of an individual learner, for example by extracting information on how their pace fluctuated across the length of the project. While we did use the data on an individual basis to create specific case studies, by and large we wanted to be able to identify the bigger pictures and trends. The 'raw data' spreadsheet of all 80 learners was too cumbersome for this and so we compressed the data in two particular ways.

First we took an average for each learner of the incidence of all coded items across the whole project. This gave us, for example, an instance of a learner having spent 16% of their time dealing with *task* issues. However, this average (as averages tend to do) masked the 'flow' of dealing with task issues at different stages of the project and so, next, we divided each project into 5 equal phases, giving us data on the first 20% of time, the second 20% and so on. This allowed us to see, for example, that the learner who had spent 16% of time across the project dealing with task issues had focussed on these in very different amounts as the project progressed, dealing with them in the following proportions:

    Phase 1   27%
    Phase 2   12%
    Phase 3   2%
    Phase 4   24%
    Phase 5   15%

Because the projects varied so much in length, by compressing the data in this way we could make comparisons across projects – for example, on the variations in the percentage of time spent on evaluating in phase 1. We could further compress

145

the data to the level of the school, the Year group or the Key Stage (phase of schooling). We called this compressed data *datamaps* as, at a literal level, they allowed us to see the terrain of the data and to see trends and contrasts in a way that the full spreadsheets denied. Having created the datamaps, we could then mine the data in a range of ways, such as:
- creating broad overviews – mapping the terrain;
- looking for patterns and follow leads;
- answering specific questions.

To illustrate our approach and the potential of the data, we give the following examples.

Mapping the terrain

Viewing the analysis of the compressed pre-coded data enabled us to provide overviews across the whole age range surveyed in ways that we do not believe had been done before. For example, using data compressed into the 5 'phases' of the projects and the 4 Key Stages of our National Curriculum (KS1 5-7 year olds; KS2 7-11 year olds; KS3 11 – 14 year olds; KS4 14-16 year olds) we were able to show the way pace varied. Figure 7 shows the percentage of time spent 'motoring' by each of these groups across the life of their projects.

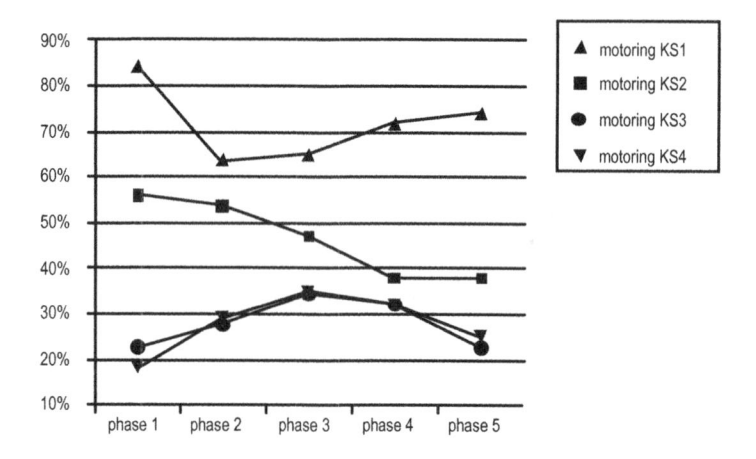

Figure 7 'Motoring' across the 4 Key Stages

This data not only showed the vast difference in engagement (and energy levels!) of the younger and older learners, but also began to indicate some distinct differences in profile – for example the difference in the general trend of primary and secondary, providing us with a pointer to explore further.

Looking for patterns and following leads

From profiles of Key Stage performance, we went on to explore the way the data presented composite Year groups (from a range of schools). The following example considers the amount of time learners spent being directed by the teacher and the amount of time being supported. Figure 8 shows the broadly similar patterns across all primary Year groups and those of Key Stage 4, and the wildly contrasting picture provided for Year 7 and Year 8.

Levels of teacher direction and teacher support across all year groups

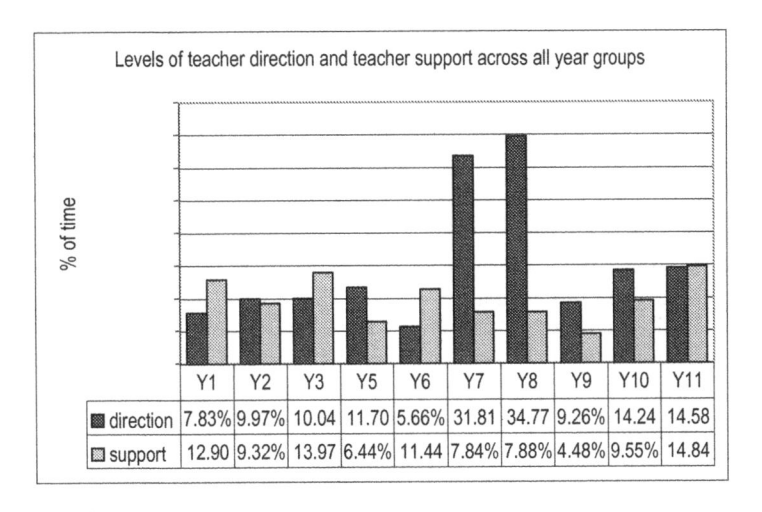

	Y1	Y2	Y3	Y5	Y6	Y7	Y8	Y9	Y10	Y11
direction	7.83%	9.97%	10.04	11.70	5.66%	31.81	34.77	9.26%	14.24	14.58
support	12.90	9.32%	13.97	6.44%	11.44	7.84%	7.88%	4.48%	9.55%	14.84

Figure 8 Direction and Support across year groups

This data raised questions about the early years of secondary schooling, and particularly the impact transition to secondary schooling appeared to have on learner autonomy. To follow up on this we turned to individual case studies, working initially from the narrative account of a Year 7 project where the pedagogy of the teacher appeared to be encouraging learner autonomy. By charting the 'direction' and 'support' data 5 minutes by 5 minutes across the length of the project it is possible to see how the learner autonomy did actually grow, as the balance shifted from teacher direction to teacher support.

Teacher predominantly in control ▢
Pupil predominantly in control ▬

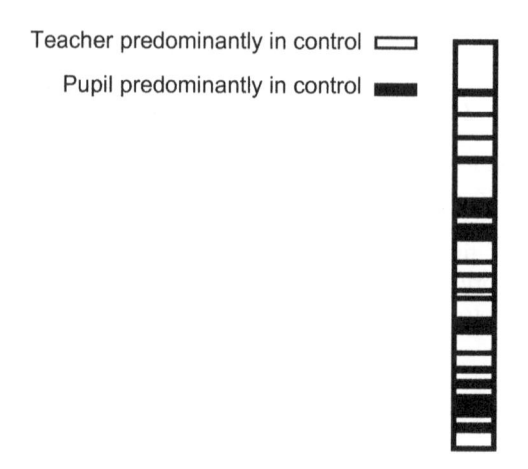

Figure 9 the growth in autonomy of a Year 7 learner

Answering specific questions

A major research question for this project related to the different ways in which learners approached design and technology and so we interrogated the data to answer specific questions in this area. For example, we wanted to know how different age groups generated ideas. To explore this, we analysed the relationship between the design intention of generating ideas and the ways this was manifest – with fascinating results, indicating somewhat surprisingly that the children in the early stages of primary school engaged in a far wider range of ways of generating ideas than those in the early stages of secondary school. As can be seen in figure 10, when the youngest learners were generating ideas they did so through 2D, 3D and discussion. The older learners worked predominantly in 2D. (Fuller accounts of the outcomes of the research can be found elsewhere e.g. Kimbell et al, 1994; Stables, 1995; Kimbell et al, 1996; Stables & Kimbell 2006; Kimbell & Stables, 2006).

CLASSROOM OBSERVATIONS

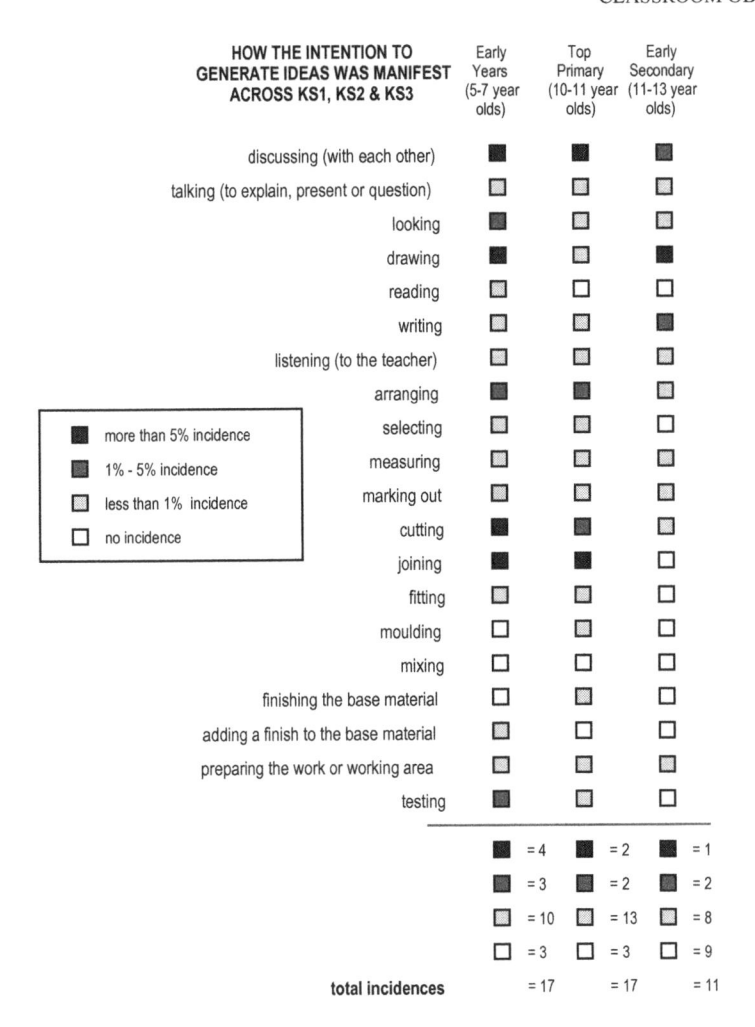

HOW THE INTENTION TO GENERATE IDEAS WAS MANIFEST ACROSS KS1, KS2 & KS3	Early Years (5-7 year olds)	Top Primary (10-11 year olds)	Early Secondary (11-13 year olds)
discussing (with each other)	■	■	■
talking (to explain, present or question)	□	□	□
looking	■	▨	▨
drawing	■	▨	■
reading	□	□	□
writing	□	▨	■
listening (to the teacher)	▨	▨	▨
arranging	■	■	▨
selecting	▨	▨	▨
measuring	▨	▨	▨
marking out	▨	▨	▨
cutting	■	■	▨
joining	■	■	□
fitting	▨	▨	▨
moulding	□	▨	□
mixing	□	□	□
finishing the base material	□	▨	▨
adding a finish to the base material	▨	□	□
preparing the work or working area	▨	▨	▨
testing	■	▨	▨

Legend:
- ■ more than 5% incidence
- ▨ 1% - 5% incidence
- ▨ less than 1% incidence
- □ no incidence

	Early Years	Top Primary	Early Secondary
■	= 4	= 2	= 1
▨	= 3	= 2	= 2
▨	= 10	= 13	= 8
□	= 3	= 3	= 9
total incidences	= 17	= 17	= 11

Figure 10 Manifestations of generating ideas in different age groups

METHODOLOGICAL STANCE

Looking back over the approaches taken to observation, to the data collected, and to the findings produced it is clear that this research didn't fall neatly into any particular research 'camp'. Viewed schematically it can be seen that we moved between what might be seen as a more positivist approach regarding the observation, to a more interpretivist approach with the findings.

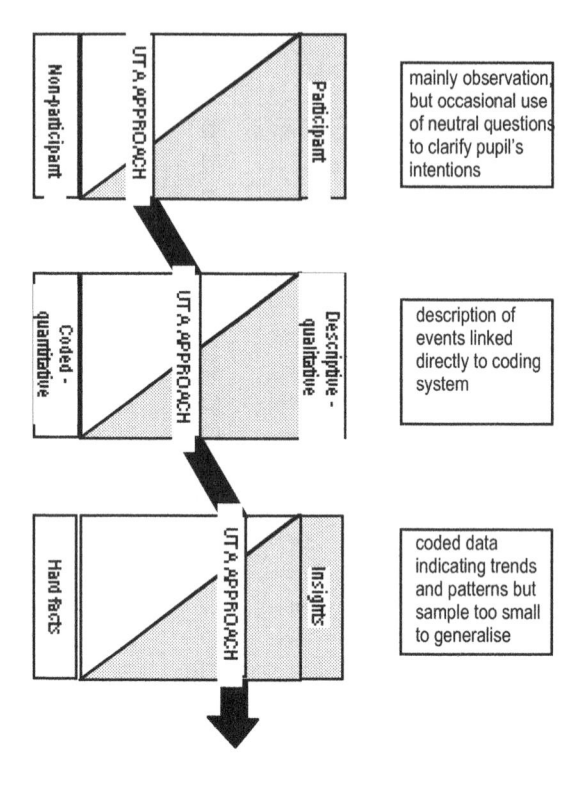

Figure 11 Schematic of methodological approaches

In reality we were almost certainly operating in an increasingly common manner of social science research, which draws on a range of methods, not at random, but in an intelligent way to achieve the job in hand. To do this we were balancing a commitment to understanding the nature of design and technological activity and the development of capability, a commitment to undertake the research with as much rigour as possible and a pragmatism driven by the resources available. As Freebody points out, the quality of research is not determined by the method itself, but by the way the method is used.

> The fundamental ways to enhance reliability and validity are the same for ethnographic research as they are for any other kind – through ensuring the clarity and accuracy of the representations of: the context of the research; the statement of the problem to be investigated; the ways in which the researcher gained access to the data; the assumptions of the participants; and the understandings on the site about the researcher's role as researcher. (Freebody, 2003, 77)

ADOPTION AND ADAPTATION OF THE METHOD

We consider that the observation framework has much potential for further use – either by adopting or adapting its structure, content and protocols. Initial subsequent use and development was by two Goldsmiths PhD students. Olefile Molwane (Molwane 2002) was undertaking research into assessment approaches in Botswana classrooms and was interested to explore the day-to-day reality of technology education in those classrooms. For his research the framework was utilised with only minor adaptations.

However, Horne (1999) was interested in what teachers did in classrooms (rather than learners) and explicitly how they made use of the physical space. Horne was keen to explore *what* teachers did in different classrooms and different subjects (not just technology education), *where* in the classroom they did it – and *why*. She took the broad principle of our observation framework and adapted it considerably to suit her own research purposes. First she created a system for coding where in the classroom the teacher was at any given moment. She did this by creating a grid, based on metre squares, of each classroom observed, which showed the location of all fixed item and furniture. She then plotted the movement of the teacher every two minutes of the lesson. Next, through exploratory trials, she created the 'what' of their actions – the equivalent of our 'manifestations' coding. She constructed a list of all observable actions teachers undertook in classrooms that could be used to create coded data of the teacher's actions. Finally she recorded the 'why' of their actions – effectively her speculations on why they were doing what they were doing. Unlike the pre-coded approach we took to learner intentions, Horne recorded her speculations within a narrative and then used this as the basis for discussions with the teacher once the lesson was concluded. So taking the broad structure and principles of our approach, she created highly detailed maps of the uses that teachers made of classrooms.

We have used the framework in two subsequent projects within TERU – in both instances for slightly modified purposes from the original. The first usage was in a project where we were evaluating a joint curriculum initiative between LEGO Education and the UK National Endowment for Science, Technology and the Arts (NESTA): ' Energy and the Environment'. Our evaluation (Kimbell et al., 2002) included a requirement to see how the classroom resources produced for the curriculum initiative were used in practice. The schools involved in the project were spread across the UK and the research resources did not enable full observation of whole projects. But we could make good use of the observation framework as a 'dipstick' – taking it into classrooms across the full sample of schools to observe particular lessons. While not providing the level of detail on individual learners that the original project had created, we were able to derive comparative data of classroom use of the resources being evaluated.

Most recently we have used the framework in the Assessing Design Innovation project (Kimbell et al., 2004) where, in the early stages, we were collecting data on the pedagogies used by 'best practice' teachers to support creativity and innovation in design and technology projects. To do this we asked the teachers to run a two-day design project with learners that was focused on creative and innovative ideas and we used a modified version of the framework to record the pedagogic 'tricks' these teachers employed and how the learners reacted. The results from these observations supported two developments for the main fieldwork of the project: first the development of assessment activities that would generate evidence of creativity and innovation; and second the creation of the assessment rubric used to make judgements on the quality of the evidence created.

ALTERNATIVES NOT PURSUED

In describing what we did do in the research it is probably worth adding a brief note about what we didn't do – but might have done, in undertaking to observe design and technological activity.

First, we didn't use video. Our main reason for not doing so was that, while it would have provided a record of the activity that could have been analysed in far greater detail, and by different teams of people, it would have eaten into our resources such that we couldn't have conceivably observed the cumulative number of hours of learner activity achieved through this project –limiting observation of either the range of learners or the range of activity, or both.

Second we didn't use the more explicit, formal methodology of protocol analysis, whereby the 'actor' literally verbalises their thoughts and actions as they are working in what is referred to as 'concurrent verbalisation (Cross et al, 1996). This methodology has been used to some effect in the professional design world (op. cit.) but for our purposes would have been too intrusive. However, in constructing the *manifestations* coding we did witness 'thinking aloud' and so included it in the list, but the only learners ever to operate in this way were 6 year olds, who, developmentally, can pass through a phase where actions are, almost sub consciously, verbalised in this way.

While these approaches weren't right for our research, they would be worth considering for others about to engage in close observation of activity.

THE IMPORTANCE OF SEEKING NEW AND DIFFERENT APPROACHES

The methodology described in this chapter, while drawing on certain traditional approaches, was derived empirically, in response to a particular and specific research need to understand processes and approaches in technology education better . It is widely accepted that, internationally, research in this area is in an early stage of development – compared with other areas of education. But, as the presence of this book testifies, it is a growing area. While not wishing 'special pleading', the procedural nature of the subject, along with the diverse roots of its development demand that new approaches are developed – be they adaptations or entirely novel approaches. It is important that researchers in technology education feel able to act creatively in deriving new approaches, not to be trapped inside this or that research paradigm. In moving the field forward it is worth referring back to

Freebody (quoted earlier in this text) who made the point that it is not the method itself that is important, but the way in which the method is used, and to reflect on the good sense of Eisner, who claimed

> Because I am a conceptual pluralist, I believe it is important from an epistemological perspective for scholars to have available to them different methods for the study of education. Different methods make different forms of understanding possible. Hence, I am seeking neither a new hegemony nor a new orthodoxy, but rather the expansion of the utensils in our methodological pantry. (Eisner, 1993, 54-55)

Eisner is highlighting the critical importance of understanding - and what is research for if not to increase understanding. Whatever the method used, whether adopted, adapted or newly created, the driving intention must be the understandings sought. At the end of the day, the method is a tool, a means to an end. Hopefully the tools described here will be amongst other new approaches added to the 'methodological pantry' that will facilitate new understandings in Technology Education.

REFERENCES

Buchanan, R. (1995). Wicked problems in design thinking. In V. Margolin & R. Buchanan (Eds.), *The idea of design* (pp. 3-20). Cambridge, Massachusetts; London, England: MIT Press.

Cross, N., Christiaans, H., & Dorst, K. (Eds.). (1996). *Analysing design activity*. London: Wiley.

Denscombe, M. (2003). *The good research guide for small-scale social research projects* (2nd ed.). Maidenhead, UK: Open University Press.

Eisner, E. (1993). The emergence of new paradigms for educational research. *Art Education, 46*(6).

Freebody, P. (2003). Qualitative *research in education: Interaction and practice* (2004 re-print ed.). London: Sage.

Gorard, S., & Taylor, C. (2004). *Combining methods in educational and social research*. Maidenhead: Open University Press.

Horne, S. (1999) *The classroom environment and its effects on the practice of teachers*. University of London, London.

Kelly, A. (2003). Research as Design. *Educational Researcher, 31*(1), 3-4.

Kimbell, R., Stables, K., Wheeler, T., Wozniak, A., & Kelly, A. V. (1991). *The assessment of performance in design and technology*. London: SEAC / HMSO.

Kimbell, R., Stables, K., & Green, R. (1994). Understanding technological approaches: *Final Project Report to ESRC*. London: Goldsmiths University of London.

Kimbell, R., Stables, K., & Green, R. (1996). *Understanding practice in design and technology*. Buckingham UK: Open University Press.

Kimbell, R., Balchin, T., & Stables, K. (2002). *Energy and the environment: An evaluation of a NESTA/LEGO collaborative design and technology project*. London: Goldsmiths University of London.

Kimbell, R., Miller, S., Bain, J., Wright, R., Wheeler, T., & Stables, K. (2004). *Assessing design innovation: A research and development project for the Department for Education & Skills (DfES) and the Qualifications and Curriculum Authority (QCA*. London: Goldsmiths University of London.

Kimbell, R., & Stables, K. (2007). *Researching design learning*: Springer.

Molwane, O. (2002) *Design and technology in Botswana: Teachers' practices and the assessment of student performance*. University of London, London.

Seale, C. (2004). Coding and analysing data. In C. Seale (Ed.), *Researching society and culture* (2nd ed., pp. 306 - 321). London: Sage.

Stables, K. (1995). Discontinuity in transition: Pupils' experience of technology in Year 6 and Year 7. *International Journal of Technology and Design Education, 5*, 157-169.

Stables, K., & Kimbell, R. (2006). Unorthodox methodologies: Approaches to understanding design and technology. In M. J. d. Vries & I. Mottier (Eds.), *International handbook of technology education: Reviewing the past twenty years*. Rotterdam: Sense Publishers.

Kay Stables
Technology Education Research Unit
Goldsmith's College
London University
United Kingdom

JOHN STEVENSON

CAPTURING KNOWLEDGE AND ACTIVITY

ABSTRACT

Because of the contextualisation of the knowledge involved in activity, judgements have to be made about the generality or specificity of any descriptions of the knowledge that are generated. This chapter outlines a research study that sought to capture the knowledge needed for effective activity in a workplace setting and the problems of attributing knowledge to the activities. The chapter focuses particularly on the knowledge of technology needed in this work, the methods employed to elicit information about this knowledge, the frameworks that could be superimposed on descriptions of that activity and the challenges that need to be overcome in delineating the knowledge that is involved. Hence the study has implications for one of the central problems in education, viz. in identifying the knowledge that should be the subject of instruction. That is, in designing instruction to prepare learners for engagement with technology (as for any learning activity), curriculum development has to infer the knowledge that is involved in effective relevant activity. The identification and description of this knowledge is an inferential process, and therefore gives rise to questions of validity.

INTRODUCTION

In this paper, it is assumed that learning is an activity. Hence, as with other kinds of activities, someone has to choose a focus for the learning activity. In formal instruction, the focus is usually on selecting learning processes, which usually consist of activities and topics, chosen in order to meet specified or implicit learning objectives. These activities, together with the relevant topics, objectives and outcomes are usually taken to be the content of instruction. Historically, there has been a tension between pre-specifying outcomes and processes of instruction (e.g. Mager, 1962; Stenhouse, 1978). In contemporary educational settings, there continues to be various ways in which curricula are conceptualised including emphases on both outcomes (e.g. competence) (see, for instance, Mayer, 1992) and processes (e.g. problem-based learning). In technology education itself, there has historically also been an emphasis on processes (e.g. design-make-appraise) (Curriculum Corporation, 1994) based on the view that a "knowledge" of technology involves a knowledge of materials, processes, systems and information; of how to engage in a problem-solving process involving the adequate construction of the problem and the fashioning of satisficing solutions (Middleton 1998, 2002); and of how to undertake an evaluation of the effectiveness of the solution.

H.E. Middleton (ed.), Researching Technology Education, 155–171.
© 2008 Sense Publishers. All rights reserved.

There are of course many assumptions involved in constructing curricula on any of these bases. The major assumption that is the focus of this chapter is on the nature of technology knowledge, its level of generality and the ways in which it is accessed in problem-solving activity. This chapter examines the problem of undertaking research aimed at determining the "knowledge" that should be the target of instruction. How can this knowledge be captured and what are the implications for teaching? The research that is critiqued is a study (Beven, 1996, 2002), which was aimed at determining the extent to which the description of the key competency "using technology" (Mayer, 1992) was an adequate representation of the knowledge of operators using information technology in the workplace. The Beven study was part of a wider study (Stevenson, 1996) and the research methods were applied across the entire study. While it is recognised that technology educators do not seek to prepare learners directly for the workplace, educators are interested in the knowledge needed in work, connections between technology knowledge and technological literacy in the workplace and the potential for transfer of technology knowledge to specific applications. So Beven's research on the workplace meanings of technology is interesting for its contribution to understanding the nature of technology knowledge that educators seek to develop.

In the following sections, the research methods of the wider study within which the Beven research was conducted are firstly outlined. This includes an overview of the ways in which the "knowledge" was captured. Then the problem of describing this knowledge is examined. This includes the difficulties of organising and grouping the activities that were observed and spoken about, and selecting verbal labels to describe the activities and/or the underlying knowledge, while still preserving the contextualised nature of what the operators actually knew i.e. their own personal meanings. The meanings that the operators appeared to have constructed are contrasted with those of the taxonomies that are usually constructed to describe this knowledge. Finally, the implications for future research are discussed.

RESEARCH METHOD

Overview

Beven's research was of front office staff working in motels in various locations in Australia. These staff interacted with customers at the front desk handling reservations, arrivals, accounts, departures and so on. In doing so, they usually interacted with a database to record, retrieve and manage information. The study was part of a wider study (Stevenson, 1996) to examine the extent to which workplace competences were generic. Four of the key competences / core skills usually identified for educational attention (e.g. see Mayer, 1992; Secretaries of State for Education and Science, England and Wales, 1991; US Department of Labour Secretary's Commission on Achieving Necessary Skills, 1992) were studied, viz.: literacy and communication, numeracy, use of computer software and complex problem solving. In this chapter it is the research on the use of computer software that is examined, particularly, but the discussion applies equally to the other areas of investigation in the wider study.

Approach

The method for the wider study was essentially a series of case studies involving direct observation and recording of workplace activity. It was a "bottom-up" analysis of workplace practices (Beven & Duggan, 1996, 12).

Beven and Duggan (1996, 12-13) introduce the method as follows.

In this research, observations of practice are supplemented by practitioners' accounts of their own practice and experience. The purpose of these accounts is to provide the experts' direct interpretation of activities observed, and insights regarding the development of the skills and knowledges they apply in vocational settings. These accounts also provide information regarding practices not captured during the observation sessions. Methodological issues pertaining to interpretive accounts of practice - such as critical discrepancies between reported thought and observed action - are acknowledged. For the purposes of this research, however, the significance of these factors is mitigated by the direct observational data with which interpretive accounts are cross-examined.

To capture a detailed record of vocational practices and the contexts in which these practices were applied, a largely ethnographic approach was used for data collection. Observations of front office / reception activities were recorded and artefacts of resources used in these activities were collected. The observational recordings were then analysed and disaggregated into individual activities. Once activities were separated out in this manner it was possible to identify individual actions which comprised these activities and to examine relationships between actions and the various contexts in which they occurred. Using this process it was possible to deconstruct contexts of practice, and to ascertain where skills were generic and the ways in which other skills were context dependent.

Beven and Duggan relate this approach to the advice of Scribner (1983):

If skill systems are active or practice-dependent one way to determine their characteristics and course acquisition is to study them as they function in these practices ... the practices themselves need to become the object of the study (Scribner, 1983, 7).

Thus, the research sought to capture activity using a variety of methods and to relate the activities to verbal descriptions of competence.

Procedure

The primary data for the wider study consisted in records of daylong observations of staff performing their normal work; recording use of computer software; and semi-structured interviews (including stimulated recall), as follows.

Observations were video-recorded using a camera set up in the reception area, complemented with an audio recording to ensure that sound was not lost. The observations concentrated on the activities of an experienced front office staff member interacting with customers and artefacts in the setting, and, where possible, interactions with less experienced staff. During the front office observation sessions, the staff described and explained activities as they undertook them. They also described and explained the resources they used, as well as the range of tasks and interactions required of them in their overall work. Data from video-recordings and interviews were transcribed into a written form.

In addition, for the Beven study, particularly, the use of computer software was recorded directly from the computer terminal to videotape, at sites where this was technically possible. This was achieved by means of a conversion device (Genlock) that was connected directly to the computer system and which converted signals from the computer image into a standard television PAL format able to be recorded on a video-recorder.

The purpose of supplementing observations with interviews was to record verbal accounts of workplace practices. Interviews were conducted with the staff members who participated in observational sessions, persons responsible for staff training, and the site manager and/or any persons with responsibilities for overall management of the operation. Semi-structured interviews were conducted with front office staff after the observation sessions. These interviews involved stimulated recall exercises regarding activities performed during the observation sessions. The front office staff members also discussed their perspectives on vocational training and reflected on their own experiences in developing vocational skills. Interviews with back office/accounting staff focused on the practitioner's use of computer software. Discussions with managers and owner-operators ranged over problem solving skills, staff training and discussion of contextual/operational features of the business. All interviews and discussions with participants were audio-recorded.

INTERPRETING THE FINDINGS

A Cultural Historical Activity Theory framework is used as a theoretical framework in the discussion of the research, its methods and its findings, in this chapter. The framework utilises Leont'ev's (1981) ideas of activity, action and operation as follows:

> When, for example, a learner hits a target in shooting practice, he [sic] performs a definite action. What characterises it? First of all, of course, what the activity is that it forms part of, i.e. what its motive is, and consequently what sense it has for him. But it is also characterised by something else,

namely by means of the operations by which it is performed. The aiming of the shot calls for many operations, each one corresponding to certain conditions of the action: it is necessary to put the body in a certain position, to align the sights of the rifle and correctly establish the line of sight, to press the butt to the shoulder, to hold the breath, and to press the trigger smoothly. For the trained marksman none of these processes is an independent action, and their objectives are not singled out in his consciousness... There is only one aim, to hit the target; and that means that he has command of the motor operations necessary for shooting. (Leont'ev, 1981, 235).

Thus, for Leont'ev, there is a hierarchy among activity, action and operation. In the motel front office, activity is the overall collective work directed at some value-laden object such as the success of the business (Stevenson, 2002a). The actions are the individual contributions to this overall object. For instance, checking-in a customer makes little sense on its own. It makes sense because of the way it contributes to the overall purpose of the activity. Operations are the individual components of the individual actions – in this case, greeting the customer, the keystrokes involved in entering and recording the information correctly, issuing the key and so on.

The framework also adopts Vygotsky's (1986, originally written in 1934) differentiation of sense, meaning and thought.

> Our investigation established three main semantic peculiarities of inner speech. The first and basic one is the preponderance of the *sense [smysl]* of a word over its *meaning [znachenie]* – a distinction we owe to Frederic Paulhan. The sense of a word, according to him, is the sum of all the psychological events aroused in our consciousness by the word. It is a dynamic, fluid, complex whole, which has several zones of unequal stability. Meaning is only one of the zones of sense, the most stable and precise zone. A word acquires its sense from the context in which it appears; in different contexts, it changes its sense. Meaning remains stable throughout the changes of sense. The dictionary meaning of a word is no more than a stone in the edifice of sense, no more than a potentiality that finds diversified realization in speech.... A word in a context means both more and less than the same word in isolation: more, because it acquires more context; less because its meaning is limited and narrowed by the context. The sense of a word, says Paulhan, is a complex, mobile, protean phenomenon; it changes in different minds and situations and is almost unlimited. A word derives its sense from the sentence, which in turn derives its sense from the paragraph, the paragraph from the book, the book from all the works of the author (244, 245)

Thus, for Vygotsky, sense is personal: the ways that the individual apprehends experience, including feelings. Meanings are the codified representations (usually in language) of that experience, which are used collectively to communicate about experience.

The theoretical framework also adopts Leont'ev's (1981) identification of the tensions between sense and meaning.

> The discrepancy originally arising between the human group's relations to the reality around it, on the one hand, which is generalised in a system of linguistic meanings; and the personal relations of individual people, on the other hand, which form the sense of what is reflected for them, already complicates the process of awareness." (Leont'ev, 1981, 258)

These differentiations are also sourced in Vygotsky's work. For instance, writing on the relationship between words and thought, Vygotsky argued:

> But while in external speech thought is embodied in words, in inner speech words die as they bring forth thought. Inner speech is to a large extent thinking in pure meanings. It is a dynamic, shifting, unstable thing, fluttering between word and thought, the two more or less stable, more or less firmly delineated components of verbal thought. It's true nature and place can be understood only after examining the next plane of verbal thought, the one still more inward than inner speech. That plane is thought itself. As we have said, every thought creates a connection, fulfils a function, solves a problem. The flow of thought is not accompanied by a simultaneous unfolding of speech. The two processes are not identical, and there is no rigid correspondence between the units of thought and speech (249)

> Precisely because thought does not have its automatic counterpart in words, the transition from thought to words leads through meaning. (251)

These differences among thoughts and words, the differences between sense and meaning, and the internal and external ways in which thought is mediated, were seen by Vygotsky to present difficulties in communication and understanding among people.

> The problem is that thought is mediated by signs externally, but it also is mediated internally, this time by word meanings. Direct communication between minds is impossible, not only physically but psychologically. Communication can be achieved only in a roundabout way. Thought must first pass through meanings and only then through words. (252)

> To understand another's speech, it is not sufficient to understand his words – we must also understand his thought. But even that is not enough – we must

know its motivation. No psychological analysis of an utterance is complete until that plane is reached (252)

For these reasons, the discussion of Beven's findings is organised as follows in the following sections. Firstly, the actions themselves are analysed by seeking to label them and place them in groups. This process involves using words, i.e. verbal labels. Firstly, an attempt is made to use the words that staff themselves used, i.e. words for which they presumably already had meaning and where these meanings helped them to make sense of their activity. Then the actions are considered in terms of other kinds of organisations of activity taken from other communities of practice. Firstly words that have been used to cluster and describe these kinds of activities across the whole of the hospitality industry are compared with the words derived directly from the workplaces. Then words used to convey meanings in the computer world are examined.

Description and Classification in terms of Actions and Artefacts

Beven (1996, 2002) found that he could classify the kinds of computer-based actions that he observed in the Front Offices of all of the motels, as listed below:
– Creating a new reservation,
– Creating a walk-up Check-in,
– Updating a Check-in,
– Updating a reservation,
– Processing a Check-out,
– Processing Phone call charges,
– Processing Breakfast charges,
– Cancelling a reservation,
– Cancelling charges,
– Saving data,
– Sorting data.

He compiled this list from the Genlock-mediated records of computer strokes, video-recordings of activities and audio-recordings of interviews. This involved an actions-based clustering of different kinds of activities according to the inferred goals of those actions. That is, the groups and verbal labels given to them were derived from the terminology used by the operators themselves and from the structure of the databases with which they were interacting. It is the discourse of the workplace. The words are those the operators used to describe their work and that were used in the database artefact. They are words about what was being done; not words that seek to convey notions of supposed underlying knowledge. So, while not the actions themselves, they are words about actions.

In examining Action (b) further, Beven (2002) found that he could disaggregate this action into 21 different sets of operations, (Columns 1 and 2 in Table 1), derived again from the words used in the database artefact and the workplace discourse. He then classified each operation at each of three motel sites according to the ways in which the database was engaged with, in order to undertake the operations, which he called Data Entry Strategies (Columns 3, 4, and 5 in Table 1).

Table 1. Data entry strategies for the action of check-in across sites (modified from Beven, 2002)

Operation	Description	Data Entry Strategy employed		
		Site A	Site C	Site D
1	Check Availability	OI	ES	MS
2	Enter Start and End Dates	SD	SD	SD
3	Read Available Screen Map	LT	LT	LT
4	Update Reservation	OI	MS	MS
5	Enter New Reservation	OI	OI	OI
6	Enter Start and End Dates	SD	SD	SD
7	Enter Room Type	MS	LT	OI
8	Check Available Screen Map	LT	LT	OI
9	Enter Rate Code	MS	MS	MS
10	Enter Charge Type	MS	MS	OI
11	Create Guest Account	OI	SD	OI
12a	Select Guest from table	LT	LT	LT
b	or Enter new data	SD	SD	SD
13	Enter Payment Mode	MS	SD	MS
14	Enter Card Details	SD	SD	SD
15	Enter Special Arrangements	SD	LT	LT
16	Enter Business Source	MS	MS	MS
17	Enter guest Geographic Source	MS	---	MS
18	Enter Notepad Details	SD	SD	SD
19	Create Folio	SD	OI	OI
20	Enter Special Requirements	OI	SD	LT
21	Confirm Check-in	OI	OI	OI

The five different Data Entry Strategies (MS: Memory / Synonym, OI: On-Screen Instruction, LT: Look-Up Table, SD: Source Data, ES: External Source) are derived directly from the structures of the database artefact and are explained in Table 2.

Undertaking these sets of operations, using the strategies that were specific to each site for the software that was being used, made sense to the people working in each particular office – they knew what they were doing, could see how it contributed to the collective work of the group and directed their actions at the collective purpose and they undertook the work with efficiency. They could talk about the activity in the kinds of terms in (a) to (k) above. These words had meaning for them. They engaged in the activity confidently and could discuss it with the researchers.

Table 2. Description of data entry strategies (from Beven, 2002)

Strategy	Description
Memory / Synonym (MS)	The user is required to recall from memory a synonym or code to complete an entry (e.g. Enter the synonym SR for single room or a number, 1 for example, to indicate a single room)
On-Screen Instruction (OI)	The user is required to select data from an on-screen instruction (e.g. Press the F5 key to indicate single room type)
Look-up Table (LT)	The user accesses a look-up table to select the appropriate data (e.g. Pressing a function key overlays the screen with a table of data options from which to select: 　　Single Room 　　Twin Room 　　Double Room 　　Executive Suite 　　Self-contained Unit The use of directional (arrow) keys highlights a selection and pressing the enter key makes the selection and transfers the choice to input field.
Source Data (SD)	The data are sourced from a variety of documents or verbal interaction with guests (e.g. from a faxed reservation inquiry, from a walk-in guest at the front counter).
External Source (ES)	The data are not maintained by the software (e.g. a paper based 'week at site' room schedule kept manually).

However, from Table 1, when classifying operations in terms of what was being done (in terms of the kinds of strategies dictated by the software artefacts), there was considerable variation across sites. That is, it is unlikely that being able to complete an action mediated by a set of software-specific operations at Site A would equip one with being able to accomplish the same action at Sites B and C. It is for this reason that researchers seek to generate more 'generic' descriptions of actions as discussed in the following section. The problem, though, is that the words that are used do not necessarily have meaning in individual workplaces.

More 'generic' description and classification based on workplace actions

A challenge to an educator or curriculum developer is to describe activities and contributing actions in terms that promote their teaching. For instance, in seeking to guide vocational curriculum development a set of Competency Training Modules has been developed for the Hospitality Industry (Tourism training Australia & ACTRAC, 1995). It seeks to be "generic", across worksites, while still

referring to actions rather than imputing underlying knowledge. As Beven (1996) notes, competences have been included which refer to the use of reservation software in their modules, such as the following:
- Handle reservation requests
- Record reservations
- Prepare occupancy reports
- Handle group and tour bookings
- Receive deposits and prepayments
- Register guests
- Finalise guest account
- Present guest account
- Collect payment
- Make forward reservations
- Handle messages
- Post transactions and journals to correct accounts

These codifications are a little further removed from those derived directly from particular worksites as in Beven's research. They seek to be more generic, applicable to different worksites with different artefacts. For this reason, while some might be directly meaningful to workplace operators in the terminology used, others may need to be bridged with a link between the terminologies. At the same time, it is important to note that they still relate to activity, rather than supposed underlying knowledge. On the other hand, they do involve the observer in constructing new ways of rendering the meanings that are involved: new words from other communities of practice.

Based on Vygotsky's arguments, these new words can come to mediate thinking in these workplaces only once these concepts have come into interaction with the spontaneous concepts of the workplace. It is only then that sense and meaning can be reconciled; and then words used to mediate thinking.

Description and classification based on supposed underlying knowledge

An alternative approach to those based directly on observed activity would be one which sought to identify the capacities that are supposed to underlie such competence; in order to ensure that what is taught can be related to the broader worlds of work and living. That is, the challenge would be to describe the 'knowledge' that is involved.

However, in both the case of codification as competences and when we, as observers, try to unmask the 'knowledge' that is involved, we are introducing new words, i.e. new and different ways of rendering meaning. For instance, as Beven observes, at a first level of categorisation of the knowledge that might be involved, one might generate the following kinds of groupings:

Entering New Data Sets
 - Creating a new reservation
 - Creating a 'walk-up' Check-in
Updating Data Sets

- Processing a Check-in
- Updating a reservation
- Processing a Check-out
- Processing additional data
     (Phone calls, Breakfast)

Removing Data Sets
- Cancelling a reservation
- Cancelling charges

Saving Data Sets
- System determined

Sorting Data Sets
- System determined

At a second level of categorisation (as in the Shelly and Cashman's (1990) taxonomy of information processing skills) Beven observes that this might all be called Data Manipulation: Being able to add, change and delete (manipulate) data in a database. And this may then be constructed as a component (Row 3) of an overall depiction of information skills/competence/knowledge as shown in Table 3.

The ontological and epistemological questions, then, are:
- Do these kinds of classifications and descriptions capture the activity of the workplaces and the knowledge that is involved"?
- Are these classifications and descriptions meaningful to operators in the workplace - would they know what they meant; Do they represent the ways in which individuals constructed meaning in the working activities that were observed?
- Could the knowledge described in this way be taught?
- If it could be taught, would the learners be competent in these workplaces?
- Is the 'knowledge' denoted by these labels transferable as such between motel sites, or between motel work and other work requiring interaction with databases?

JOHN STEVENSON

Table 3. Taxonomy of the knowledge and skills identified when using computer databases in motel front office workplace practice (modified from Beven, 1996, 2002)

Kinds of Knowledge - (Shelly & Cashman capacities shown in italics)	Associated sub-sets of knowledge and skills
System Knowledge (*understand the concepts of a database*)	Have a Functional Map of the system Access to System Exit from System
System Navigation Knowledge (*understand the concepts of a database*	Using Menus Using Function Keys Using Directional Keys
Data Manipulation knowledge (*be able to create and display new records in a database*) (*be able to add, change and delete (manipulate) data in a database*)	Adding New Data Changing Data Removing Data Saving Data Sorting Data
Software Maintenance Knowledge (*be able to perform routine administrative functions on the data and the database*	Undertaking user software maintenance tasks, e.g. − System Initialisation procedures − Backup and restoration procedures − End of Period procedures
Inquiry and Reporting Knowledge (*be able to sort and report data in a database*)	Screen Based Inquiry (knowledge) − Inquiry options − Relationship between options − Mnemonics, icons, codes − Interpret business problem − Construct query Producing Reports (knowledge) − Printer operation knowledge − Printer connectivity knowledge − Operating system knowledge

My contention is that while it is possible for an observer to construct these labels, abstracted from practice, and relate them to the terminology of other discourses (e.g. the wider information technology discourse), there are many problems involved, as follows.

It is a mistake to believe that knowledge as represented by the label exists, that it is generic or that it is directly transferable. That is, it is a big leap to thinking that a person 'has data manipulation knowledge' because that person can use software in a particular motel computing system in undertaking such actions as: *Creating a new reservation, Creating a walk-up Check-in, Updating a Check-in, Updating a reservation, Processing a Check-out, Processing Phone call charges, Processing Breakfast charges, Cancelling a reservation, Cancelling charges, Saving data, Sorting data.*

It is an even bigger leap to think that the capacities that are involved in, or derived from their meaning-making (called competence or 'knowledge' or by some other term) can be used directly in another motel (especially since the study found that each motel had quite different menu structures, and different key combinations were needed for apparently similar functions; and each motel had its own culture of practice). It is also a step too far to think that the competent front office motel worker would recognise their own capacities in the terms used in codifying 'data manipulation knowledge' or as a 'key competency' or 'new basic'. And, of course, one also cannot assume that the worker could automatically and directly transfer such 'data manipulation knowledge'/'competence' to another industry or to other professions using data manipulation.

Rather, what exists is: the capacity to undertake observed actions in that activity system, mediated by the artefacts and culture of that system and in concert with the collective object of that system. This meaningfulness may well be able to be connected to other ways of rendering meaning that an observer might recognise, but there is learning involved in developing a learner's capacity to access and make meaningful those other ways of understanding. Moreover, it needs to be noted that this so-called data manipulation knowledge was integrally related to other sources of meaning in their work often differentiated into such categories as values (Stevenson, 2002), literacy (Searle 2002), numeracy (Kanes, 2002) and problem-solving (Middleton, 2002); and its recognition as a particular category of knowledge is therefore arbitrary.

Of course, the opposite is also true of 'conceptual' meanings transacted outside of such vocational pursuits. It is one thing to think that one knows conceptually what is intended by *Updating a reservation*, or even being familiar with the concepts of a database, but it is another to have the meaning that comes from successful engagement with the instruments in practice in achieving this collective purpose. And as argued throughout this paper, this is the central place of the vocational, to provide opportunities for discerning significance, developing these meanings, and using them to make other kinds of knowing meaningful.

IMPLICATIONS FOR RESEARCH METHODS

The research methods used in this study and the wider study of which it was a part had considerable strengths. These included:

- The use of case studies, which is an appropriate approach when complex interacting factors are being studied in an investigation, rather than the effects of a single factor intervention.
- The triangulation of data gathering techniques, including audio and video-recordings, interviews, stimulated recall and the capture of computer interactions. These techniques enabled the researcher to describe real actions as they took place, using the terminology that had currency in the work place.
- The study of the same kind of phenomenon across a number of sites in different locations
- The suitability of the methods for capturing activity and actions without compounding these with supposed knowledge.

Further lessons that can be learned from this study flow particularly from the last-mentioned strength.

Firstly, research methodologies in education need to be driven by a theoretical position about what constitutes knowledge and activity as well as the relationships between them. Otherwise observers may inappropriately take their codifications of actions and activity as the actions themselves or as the knowledge that enables the actions and activity. The use of Cultural Historical Activity Theory as a tool in designing a research methodology when studying knowledge and activity is especially appropriate, because of:

- The ways in which it differentiates operations, actions, activity, meaning, sense and thought
- Its recognition of the cultural situatedness of activity
- Its recognition that activity and thinking are mediated by internal and external tools and signs.

These differentiations allow the researcher to study the phenomena separately and explore their interrelationships without starting with a position that confounds them.

Secondly, the study highlights the need to devise methods that capture activity in such a way that its goal-directedness, situatedness, contextualisation, enculturatedness, indivisibility, mediation, and collectivity are not left out, as noise. In recognising the range of elements that can be identified in an activity system, mapping them and exploring the tensions among them, a more adequate explanation of the activity and the supposed underlying knowledge can be addressed. As Engeström (1999, p. 3) explains: "The object is depicted with the help of an oval indicating that object-oriented actions are always, explicitly or implicitly, characterized by ambiguity, surprise, interpretation, sense making, and potential for change". These various facets of activity and knowing also need to be captured along with the actions and operations that are present. Engeström's (1987, 1999, 2002) conceptualisation of interacting elements in an activity system provides one such framework.

Thirdly, the study highlights how important it is for a researcher to acknowledge and recognise the limitations of constructing meaning on what the researcher sees and hears: i.e. to recognise it as an observer construction, using the words that have meaning for the researcher. The relationship between the researcher's meanings and the actual sense that subjects make of their actions and activity need to be checked with the subjects. That is, the ways in which subjects understand others' verbalised constructions of their activities need to be regarded as translations, not the action, activity, sense, meaning or knowledge itself.

Fourthly, the study illustrates how care must be taken in generalising or abstracting from observations of activity and from constructions or codifications of the knowledge that is supposed to be involved. That is, groupings and descriptions of activity are imposed on the activity; relating the activity to other activities must not lose the contextualisation of the activity; and suppositions about internal processes and memory structures must not be separated from the contexts within which they have meaning.

CONCLUSIONS

Research that is aimed at capturing knowledge and activity needs to be driven by a theoretical conceptualisation of the relationships among operations, actions, activity, sense, meaning, thinking and supposed knowledge. In this way, important relationships among different ways of knowing, their connections with motivation, feeling and activity will not be lost as noise.

Researchers need to be especially careful in 'abstracting' and 'generalising' from observations of activity when grouping and describing such activity and using words to do so. The verbal labels need to have currency in the activity system under investigation and have meaning for those involved in the activity system. That is, researchers need to recognise that when witnessing activity, this is not merely the execution of some generalised concept or procedure. Rather its sense comes from the cultural context and the overall purposes of the activity.

When making a further leap in identifying the 'underlying knowledge', it is also important not to abstract from the realities and concreteness of real situations. As Van Oers (1998) argues, the idea of decontextualisation, when generalising, needs to be replaced with recognition that knowledge can only be re-contextualised. Individuals struggle to make new sense of what had sense in another context, and this is a struggle as they relate the ideas to new concrete realities. That is, Van Oers (1998) views context as providing for 'two essential processes: it supports the *particularisation of meanings* by constraining the cognitive process of meaning construction, and by eliminating ambiguities or concurrent meanings that do not seem to be adequate at a given moment; on the other hand, context also prevents this particularised meaning from being isolated as it *brings about coherence* with a larger whole (Van Oers 1998, 475).

With an appropriate theoretical conceptualisation, there are many possible research techniques that can be drawn upon when seeking to capture activity and knowledge. This study illustrates the importance of seeking to have participants render their sense and meanings in a variety of ways: through actions (captured on video-tape), through words (captured on audio-tape), interpretations (captured through interview and stimulated recall), and through interaction with artefacts (captured by converting computer signals to video signals). This kind of triangulation assists with validity, but it also brings to the fore the different ways in which actions and operations are 'understood'. Observer interpretations such as those that are imposed through other kinds of classifications of 'competence' or 'knowledge' need to be referred back to the participants in such studies to check for their validity in terms of whether the new words make sense.

Finally, a good theoretical framing in designing the study and selecting research methods becomes important again in analysing the data. In this study, using such conceptual differentiations as those of operations, actions, activity, sense, meaning and thought (from Leont'ev and Vygotsky) as well as Engeström's depiction of activity systems were important in developing an understanding of strengths and limitations of different analytical frameworks.

REFERENCES

Beven, F. (1996) Using technology. In J.C. Stevenson (Ed.), *Learning in the workplace: Tourism and hospitality*. (Chapter 6, pp. 123-144). Brisbane: Centre for Learning and Work Research, Griffith University.

Beven, F. (2002). The knowledge required for database use. *International Journal of Educational Research*, 37(1), 43-65.

Beven, F. & Duggan, L. (1996). A conceptualisation of generic skills and context-dependent knowledge and a methodology for examining practice. In J. C. Stevenson (Ed.), *Learning in the workplace, tourism and hospitality*. (Chapter 2, pp. 7-21). Brisbane: Centre for Skill Formation Research and Development, Griffith University.

Curriculum Corporation (1994). *A statement on technology for Australian schools*. Carlton, Vic: Curriculum Corporation.

Engeström, Y. (1987). *Learning by expanding: An activity-theoretical approach to developmental research*. Helsinki: Orienta-Konsultit.

Engeström, Y. (1999). Expansive learning at work: Toward an activity-theoretical reconceptualization. Keynote address, *Changing practice through research: Changing research through practice*, 7th Annual International Conference on Postcompulsory Education and Training. Centre for Learning and Work Research, Griffith University: Brisbane.

Leont'ev, A. N. (1981). *Problems of the development of the mind*. Moscow: Progress Publishers

Mager, R. F. (1962). *Preparing objectives for programmed instruction*. San Francisco: Fearon

Mayer, E. (Chair). (1992). *Key competencies*. Report of the committee to advise the Australian Education Council and Ministers of Vocational Education, Employment and Training on employment-related key competencies for postcompulsory education and training. Australian Education Council and Ministers of Vocational Education, Employment and Training.

Middleton, H. (1998). *The role of visual mental imagery in solving complex problems in design*. Unpublished PhD Dissertation. Brisbane: Griffith University.

Middleton, H. E. (2002). Complex problem solving in a workplace setting, *International Journal of Educational Research*, 37, 67-84.

Stenhouse, L. (1978). *An introduction to curriculum research and development*. London: Heinemann.

Stevenson, J.C. (Ed.). (1996). *Learning in the workplace, tourism and hospitality. A report on an initial exploratory examination of critical aspects of small businesses in the tourism and hospitality industry*. Brisbane: Centre for Skill Formation Research and Development, Griffith University.

Stevenson, J.C. (2002a). Normative nature of workplace activity and knowledge. *International Journal of Educational Research*, 37, 85-106.

US Department of Labour Secretary's Commission on Achieving Necessary Skills. (1992). *Learning a living: A blueprint for high performance*. A SCAN's report for America 2000. Washington DC: US Department of Labor.

Secretaries of State for Education and Science, Employment and Wales. (1991). *Education and training for the 21st century*. May. London: HMSO.
Tourism Training Australia & ACTRAC (1995). *Competency based training modules for the hospitality industry*. 1995 edition. ACTRAC.
Van Oers, B. (1998). From context to contextualisation. *Learning and Instruction*, 8 (6), 473-488.
Vygotsky, L. (1986) [1934]. *Thought and language*. (trans. and ed. A. Kozulin)Cambridge, England: the MIT Press

John Stevenson
Griffith Institute for Educational Research
Griffith University
Australia

BRADLEY WALMLSEY

USING STIMULATED RECALL TECHNIQUES IN TECHNOLOGY EDUCATION CLASSES

INTRODUCTION

This Chapter provides insights into why and how technology education researchers should and would seek to understand the unique and complex interactions between teachers and students within technology education classroom activities. More specifically, this chapter details the research method of video-stimulated recall interviewing. This technique was used as one component of a study into technology education learning environments to understand better how these environments might support an increase in students' use of higher-order thinking skills. A number of technology education classrooms exhibiting varying curricula and pedagogical styles were first video recorded and subsequent interview discussions with teachers and students were stimulated using the classroom video footage. The results of these interviews and video-data were used to adjust technology learning environments such that the higher-order thinking of students was promoted. This chapter provides details of data collection techniques and data analysis in gathering both the video and stimulated recall interview data within the broader context of the research study.

STUDY OVERVIEW

Introduction

The study examines issues relating to cognition, activity and the learning environment in year nine (13 to 14 year old students) technology education classrooms that involve students in designing activities that are supported by knowledge of various technological and industrial materials, processes, information and systems.

Syllabus documents (QSA, 2003), standards statements (ITEA, 2000), and curriculum documents (QCA, 1999; New Zealand Ministry of Education, 1995) indicate a new pedagogic focus for technology education. That is, workshop based subjects that have traditionally provided students with opportunities to experience industry or craft-related hand and machine skills, (Young-Hawkins & Mouzes, 1991) are gradually being superceded by technology subjects that focus progressively more on critical and creative (i.e. higher-order) thinking skills (Lee, 1996). These types of technology subjects are designed to respond to societal changes, such as those evident in many of the world's current post-industrial technological societies (Lauda, 1988).

H.E. Middleton (ed.), Researching Technology Education, 172–192.
© 2008 Sense Publishers. All rights reserved.

Technology students are now being provided with opportunities to develop complex higher-order thinking skills by "working technologically" (QSA, 2002, 4) and gaining "technological literacy" (ITEA, 1996). One of the principles that guide the formation of these curricular standards is the requirement for students to participate in "active and experiential learning" (ITEA, 2000, 3). An emphasis has been placed on the importance of combining both thinking and practical activities during a technological design process (ITEA, 1996).

Theoretical framework

Technology education's potential to support students as they 'work technologically' and gain 'technological literacy' inspires researchers to pose questions about such issues as; the nature of design and problem-solving (Lee, 1996; McCormick, 1996; McCormick & Davidson, 1996; McCormick, Murphy & Hennessy, 1994; Williams, 2000); the structure of technological knowledge (McCormick, 1997; Schultz, 2000; Herschbach, 1998, 1995; DeMiranda & Folkestad, 2000) and teaching strategies that facilitate student learning in technology education (Eggleston, 1992; Fritz, 1996; Hansen, 1995, Johnson, 1996; Kemp & Schwaller, 1988; Lauda 1988). In response to these types of questions, this study examines year nine technology classrooms, in terms of teacher and student activities. Because of this examination, the study is able to analyse teacher and student reactions and interactions during; design and problem-solving; particular teaching tactics; and different types of student thinking in technology education classes.

Elshout-Mohr, Van Hout-Wolters and Broekkamp (1999, 68) refer to three "fundamental knowledge systems" that provide the basis for teaching, "lesson structure, subject matter and learning goal". Learning goals are sought using instructional strategies. Teachers formulate instructional strategies in response to requisite subject specific learning outcomes or a process of "front-end analysis" (Jonassen, Grabinger & Harris, 1990). Jonassen, Grabinger and Harris (1990) define front-end analysis as the act of identifying the central focus of the instruction in terms of meeting the specific needs of the learner in a particular content area. In technology education, documented learning outcome statements generally stipulate this front-end analysis.

Instructional strategies that facilitate learning goals are evident at three levels. Rothwell and Kazanas (1992) and Jonassen, Grabinger & Harris, (1990) identify two distinctive levels; the macro-level and micro-level instructional strategy. Leshin, Pollock and Reigeluth (1992) identify a third intermediate-level instructional strategy that lies between the macro-level and micro-level strategies.

The macro-level instructional strategy is a general plan that defines a yearly program, unit or module of instruction and sequences a series of learning outcomes into an integrated instructional entity (Jonassen, Grabinger & Harris, 1990; Rothwell & Kazanas, 1992). An example of a macro-level instructional strategy in technology education would be a yearly program, unit or module of student technological activity focused on the systems of manufacturing, or the appropriateness of a new technology for a particular culture.

The intermediate-level instructional strategy links several areas of instruction through particular instructional approaches (e.g. discovery approach, expository approach). In technology education for example, a teacher may decide to structure a unit, lessons, lesson or mini-lesson to allow students to discover material properties (e.g. wood, metal, plastic etc.) through experimentation (e.g. destructive testing). The intermediate-level instructional strategy provides cohesion and significance for a series of micro-level instructional strategies, which when combined with other intermediate-level strategies, supports the intended instructional strategy at the macro-level (Leshin, Pollock & Reigeluth, 1992).

The micro-level instructional strategy is more specific and occurs within the level of an individual lesson or mini-lesson. It defines what students are to experience during a discrete learning event in respect to their demonstrating of a specific learning outcome (Jonassen, Grabinger & Harris, 1990). For example, a technology education syllabus document, curriculum document or standards statement identifies a set of desirable learning outcomes in a particular field (e.g. manufacturing technologies, transportation technologies, etc.). From these outcomes, a course of instruction is conceptualized and further refined into increasingly discrete units, until each student is engaged in a series of activities that supports their displaying of an intended outcome. It is during student learning activity that teachers facilitate particular outcomes using "simple instructional tactics" (Rothwell & Kazanas, 1992, 178). An instructional tactic is defined as what teachers do, (i.e. an activity or series of activities), and they are the component parts that facilitate a particular instructional strategy (Jonassen, Grabinger & Harris, 1990; Leshin, Pollock & Reigeluth, 1992; Rothwell & Kazanas, 1992).

Teaching or instructional tactics are afforded little recognition in the instructional design literature (Rothwell & Kazanas, 1992). While instructional strategies are overtly considered the central focus for many instructional designers, their interpretations of various instructional strategies commonly differ (Jonassen, Grabinger & Harris, 1990). Jonassen, Grabinger, and Harris (1990, 32) argue that;

> A strategy is like a blueprint; it shows what must be done, but does not prescribe how to do it. Instructional strategies describe a general approach to instruction but do not prescribe how to organise, sequence, or present instruction.... strategies provide useful advice about how to present and cue content, but they do not prescribe how they should be implemented. Implementations of instructional strategies are instructional tactics.

Research questions

This study examined student higher-order thinking in technology education. The study asked and subsequently sought to answer the question; how can the higher-order thinking skills of students be promoted in technology education classes? To

answer this question, four sub-questions were posed and examined during the various phases of the study. In addition, the cross-cultural aspect of this question were examined because of the study's access to year nine technology education classrooms in two countries (Australia and America).
- Phase 1. What statistical differences exist between a number of year nine technology education classes regarding the press for higher-order thinking?
- Phase 2. What factors may have an influence in regards to promoting the press for higher-order thinking within specific year nine technology education classes?
- Phase 3. Will the press for higher-order thinking in technology education learning environments be promoted through provision of information to technology educators relating to factors that are argued to promote student higher-order thinking?
- Phase 4. Do factors that are argued to promote a press for student higher-order thinking in technology education vary between different cultures?

The aim and purpose of the study

Technology education as an evolving subject area is now expected to formulate its instruction based on strategies at the macro-level that emphasise the development of students' higher-order thinking skills. The purpose of the study was to provide empirical research evidence of technology teaching and learning strategies at the micro-level that promote and improve student higher-order thinking outcomes. In addition, the study aimed to achieve this result by incorporating and valuing the existing thoughts and beliefs of teachers and students regarding the structure of current teaching and learning practices in technology education. That is, the study's intention was to refocus instruction and learning based on improving current technology classroom practices (teaching tactics) that promote higher-order thinking outcomes, rather than advocating a technology learning environment substantively alien to both teachers and students.

Importance of the study

The study's design acknowledged the lack of research that focuses on the classroom activities of teachers and students in technology education (McCormick, 1996). This study is important because it identified and subsequently sought to verify components of existing technology teaching and learning practices that were considered influential in terms of promoting students' use of higher-order thinking processes. Zohar, Degani and Vaaknin (2001, 469) state that:

> As the drive for teaching for understanding and higher order thinking gains momentum in our schools, there is a need for a deeper investigation into the conditions necessary for its success.

The study recognised that technology teachers' pedagogic philosophies are inherently stable and for many teachers these philosophies reside somewhere in the past with more traditional forms of industrial skills instruction (Shield, 1996). This study combined theory (cognition, activity and behaviour setting) and current technology teaching practices, and sought to discover that refocusing current teaching and learning strategies has potential for significant instructional reform in technology education (Zuga & Bjorkquist, 1989; Shield, 1996).

Research methodology

The study combined both quantitative and qualitative research methods to facilitate the study's objective and moved through four phases in order to answer the research questions.

Phase 1 This phase provided empirical evidence (quantitative) of the types of thinking year nine technology students are pressed to use during their learning experiences. Students were surveyed using an instrument (Cognitive holding power questionnaire, CHPQ) developed and validated to assess the extent to which the learning environment causes or presses students to use higher-order thinking (Stevenson, 1998; Stevenson & Evans 1994). This was the control phase of the study and provided empirical evidence that supported the selection of a number of technology classes for phase two.

Phase 2 This phase identified the instructional tactics (qualitative) (i.e. teacher and student actions and interactions) within a number of technology education classrooms. During this phase technology classes were video-recorded and teachers and students were subsequently interviewed using video-stimulated recall interview techniques.

Phase 3 This phase provided empirical evidence (quantitative) of the types of thinking technology students were pressed to use during this phase of the study. Teachers were advised to adjust their instructional tactics in direct reference to the evidence discovered during phase two of the study. That is, they were asked to use particular teaching tactics that appeared to be supportive of student higher-order thinking. These technology classes were pre and post-tested using the CHPQ. This formed the experimental phase of the study.

Phase 4 This phase provided supportive empirical evidence for the previous phases of the study and identified the instructional tactics of technology educators from a different culture. This phase combined both qualitative and quantitative methods.

The subsequent sections of this chapter, outline phase 2 with the aim of providing an in depth description of the video-stimulated interview research method used within this study.

PHASE 2. VIDEO-STIMULATED RECALL INTERVIEWS

Introduction

Video data collection and subsequent video-stimulated recall interviews were conducted during phase 2 of the study. This method nested within the larger study and was both supported by and supportive of the other study phases. The study combined quantitative (survey) and qualitative (case study) research methods (Babbie, 2001; Campbell & Stanley, 1963) to provide an investigative framework that facilitates the intended aim of the study.

Site and subject selection

The purposive sampling method (Bernard, 2000; Krathwohl, 1993) adopted in this study was used to ensure that the target students from each technology education class were taught similar subject content. That is, students were required to be exposed to a technology curriculum that supported their learning activities within a technological problem-solving framework through the application of propositional knowledge related to materials, processes, information and systems. It was considered almost impossible to provide a random sample of schools and obtain these subject characteristics because of the variations in current technology subject content and teaching methods (Newberry, 2001; Sanders, 2001; Fritz, 1996). The use of this sampling method does reduce the external validity (Campbell & Stanley, 1963) of the study to some extent. That is, because of the non-random nature of the sampling method employed, the results of the study can not be generalised to other technology classrooms outside the study confines. However, the changing curriculum landscape is requiring that teachers and schools change the focus of their current technology curriculum to conform to the methods and content examined in this study. Therefore, the results of this study are significant in terms of the future of curriculum development in technology education. In particular, for the development of teaching tactics that support students' use of higher-order thinking within these developing technology curriculum frameworks.

Previous research (e.g. Gump, 1978, 1991; Wicker, 1984; Schoggen & Schoggen, 1988) has highlighted the effect of school size on students' perception of the learning environment, in terms of their participation in curricular and extracurricular activities. In consideration of the possible confounding effects of school size on the results of this study, all the schools included had comparable class and total student populations. Schools were also chosen based on socioeconomic comparability. This provided the study with student populations having similar, although particular characteristics.

Approximately 1000 year nine students participating in technology education were targeted as the student subjects for this study (i.e. approximately 500 students in Australia and 500 students in America). These students were chosen in preference to older or younger students because of their recent personal choice to commence studies in the technology education subject.

Teachers agreed to participate in the study with full knowledge that the investigative processes were extensive and required a high level of commitment to the study's aims and objectives. Teachers were required to have particular attributes. For example, teachers needed to have understanding, willingness to discuss alternative teaching and learning methods, willingness to provide their time, openness to suggestions for change and a willingness to accept some minor disruptions to class routine. Some or all of these attributes could influence the external validity (Campbell & Stanley, 1963) of the study. It is argued that teachers do not readily adjust their teaching practices in recognition of research results (Elshout-Mohr, Van Hout-Wolters & Broekkamp, 1999; Slavin, 1999; Wollman-Bonilla, 2002). The methodological vigor (Levin & O'Donnell, 1999) and pragmatic nature of this current study sought to acknowledge and apply actual classroom teaching and learning practices in technology education (Wollman-Bonilla, 2002). Through this practical approach, the study aimed to support an adjustment in teaching tactics and philosophy, in order to promote student higher-order thinking in technology education classes.

Classes were selected for the second phase of the research based upon the nature of their responses to the Cognitive Holding Power Questionnaire (CHPQ) (Stevenson & Evans, 1994) administered during phase 1. It is argued that learning environments that measure higher in SOCHP (second order cognitive holding power) and lower in FOCHP (first order cognitive holding power) press students into learning situations that demand higher-order thinking skills (decision-making, critical thinking, problem-solving etc) (Stevenson & Evans, 1994; Stevenson, 1998; Walmsley, 2001, 2003, 2004). Without the student autonomy (self-regulation and effort) provided by these types of learning environments, it is not possible for students to display, or see the value of the characteristics of higher-order thinking (Perkins, 1993; Resnick, 1987). On the basis of the pre-test results, six classes (eight lessons) were chosen for further investigation, using video observations and teacher and student video-stimulated recall interview techniques (Clarke, 1998; Pirie, 1996). The criterion for choosing these classes was that they displayed statistically significant results for either FOCHP or SOCHP during the phase 1 pretest.

Video data capture method

Video Equipment A video research package (VRP) was assembled to facilitate the collection and later analysis of video data from within technology education classes during phase 2 of the study. The VRP included two digital video cameras, which captured footage of teacher and student teaching and learning activities. One camera was kept static and positioned to capture as much of the classroom activity as was possible (i.e. background activity). The other camera was used to capture the teacher's movements throughout the technology classroom (i.e. foreground activity).

The footage from the two cameras was mixed on site using a vision mixer. The vision mixer enabled the creation of a television screen view from both cameras using a picture-in-picture format. The monitor enabled the researcher to note interesting segments of the lesson for later investigation during video-stimulated recall interviews with teachers and students.

Audio was gathered via a lapel radio (RF) microphone and a shotgun directional microphone (general class). The shotgun microphone was required to permit student conversations to be monitored apart from the teacher. The mixed video and audio footage was recorded to a digital videodisc recorder on site. The recorded footage (DVD) facilitated playback of the recorded lesson using the combined functions of the DVD player and the television monitor. The researcher was then able to conduct post-lesson video-stimulated recall interviews with the teacher and select students using this facility.

Video Data Analysis The normal classroom activities of teachers and students in technology education were recorded using two video cameras. One camera focused on teacher activities and teacher to student interactions and the other on student activities and student to student interactions (Clarke, 1998). Recordings of conversations and interactions were limited to specific instances of teacher to student and student to student interaction. These instances provided opportunities to observe students and teachers as they conducted purposeful dialogue during their problem solving, questioning, explaining, monitoring and facilitating activities. Difficulties were experienced in audio recording student to student conversations due to equipment limitations and the noise levels of technology classroom environments.

The videotapes were analysed in line with the method employed by Stevenson and McKavanagh (1994). Stevenson and McKavanagh developed a coding system to analyze the activities and cognitive interactions of teachers and students within Colleges of Technical and Further Education (TAFE) learning settings. Stevenson and McKavanagh's coding system allows for analysis of the foreground activities (teacher participation) of the learning setting. The current coding system has been expanded to include a system for coding background activities (i.e. student independent of the teacher learning activity). Each code letter represents a minor code description and is contained within one of five major code categories;
– Teaching / Learning Technique
  Initiation
– Group Size
– Resources used to Aid Instruction or Learning
– Cognitive Activity.

Modifications were made to the original schedule designed by Stevenson and McKavanagh (1994) because of the specific nature of the technology education learning environments under investigation in this study. The study required the incorporation of background as well as foreground technology classroom activities within the data analysis, because of the need to identify self-directed autonomous technology learning experiences said to be necessary in the support of student higher-order thinking (Perkins, 1993; Resnick, 1987). Background activities are those that include any student learning activities that occur outside the direct influences of the teacher (i.e. in the background). Alternately, foreground activities are those that include the direct influence and attentions of the class teacher (i.e. in the foreground).

Video-stimulated recall interviews

The collection of video and interview data was subject to situation specific restrictions, in terms of class access before, during and after lessons. In addition, student interviews were restricted to around 15 minutes to allow several students' participation (Note: the number of students varied as dictated by each class activities and circumstances). The length of teacher interviews for the eight lessons averaged at around 90 minutes which is consistent with Nespor's (1984b) study. Interviews with teachers and students were audio-taped for later transcription. This ensured a minimal loss of data.

Video-stimulated recall technique (Clarke, 1998; Nespor, 1884b; Pirie, 1996) was used to focus students and teachers on specific events and activities to facilitate thoughtful and more accurate responses. Pirie, (1996) and Nespor, (1984a) suggest that these types of interviews should be held as soon as possible after the video-recorded lesson, as delays of seven days or more reduce the reliability of the data. Therefore, all interviews in this study were conducted as soon as practicable after each technology lesson (Clarke, 1998). Student interviews were conducted during the next available lesson after the videotaping and teacher interviews generally were held on the same or next teaching day. This sometimes meant that the weekend occurred between the two occasions, this factor did not appear to have a marked affect on the ability of the students or teachers to recall their thoughts within the lesson under discussion. Clarke (1998, 4) argues that;

> An individual's video-stimulated recall account will be prone to the same potential for unintentional misrepresentation and deliberate distortion that apply in any social situation in which individuals are obliged to explain their actions. A significant part of the power of video-stimulated recall resides in the juxtaposition of the interviewee's account and the video record to which it is related. Any apparent discrepancies revealed by such a comparison warrant particular scrutiny and careful interpretation by the researcher.

It is not the normal course of events for teachers firstly, to have their lessons videotaped and secondly to view and explain their thoughts and actions afterwards (Nespor, 1984b). Some teachers in this study commented that they felt uncomfortable during the process of videotaping and interviewing. The researcher conducted prior meetings with each teacher to explain the videotaping and interview process. These meetings were held in an attempt to alleviate teacher concerns. In addition, the open nature of the interview protocol permitted teachers to comment on any aspect of the lesson. In general terms, teachers considered the video-stimulated recall interview process as being a positive learning experience. Teachers also considered that they were more nervous about the process than were their students.

Calderhead (1981) argues that much of what teachers do within their classrooms becomes routine and automatic. The nature of these types of teaching behaviours is that they are often tacit and therefore difficult to vebalise and teachers may vary in their abilities to account verbally for these types of more tacit behaviours (Calderhead, 1981; Nespor, 1984b). The format of the interview process used in this study encouraged the researcher to question teachers further if their explanations of instances were incomplete or confused. This process lead to a joint construction of meaning, between the teacher and researcher, regarding situations specific to the lesson. The researcher and teacher were involved in a social event which leads to an understanding of a version of reality rather than an actual account of reality as it occurred retrospectively. Nespor (1984b, 28) argues that;

...viewing a tape of his or her classroom is viewing a different stimulus environment than the one they encountered in actually teaching the class...what the teacher sees at the end of the day on the videotape is an event about which the teacher possesses interpretive frameworks quite different from the ones he or she possessed as the class actually unfolded.

In addition, the video-footage provides the teacher with a very different view of the class than that experienced during the lesson. The various camera positions provide the teacher with access to various stimuli that may have not been previously apparent during the lesson (Nespor, 1884b). This situation may lead teachers to alter their explanations of experiences, as they happened during the lesson, to account for their subsequent observations of the lesson during the interview process. To limit this possibility, the interview protocol requested that teachers make a distinction between their thoughts as they occurred during the lesson and their thoughts about the lesson retrospectively as they viewed the video recording. Many teachers were able to make this distinction, although it is unclear to what extent this factor has influenced teacher responses. However, the video data does provide the researcher with an additional frame of reference regarding teacher responses (Clarke, 1998).

Student interviews provided a valuable means of evaluating teacher interpretations of lesson activities. Students were selected for each interview by the researcher and approved by the teacher. The reason for seeking teacher approval was to ensure that the student would be comfortable with the process and motivated to provide credible responses during the interview. Each student was approached to volunteer and several students (i.e. 3 students during 23 interviews) declined to participate in the interview process.

In spite of the various concerns highlighted above, video-stimulated recall interview technique is argued to be a useful method with which to evaluate teachers' "in-flight thoughts and decisions" (Nespor, 1984a, 120). The method provides teachers with a clear and concise picture of lesson activity. When this technique is combined with other data collection methods, a more complete understanding of classroom cognition and activity is achieved (Calderhead, 1981; Nespor, 1984a, 1984b). The interview technique used in this study provides supportive evidence for video coding categories. That is, the linkage of student and teacher dialogue with video recordings, facilitates a more complete and accurate interpretation of technology classroom events and activities (Clarke, 1998).

Video-stimulated Recall Interview Data Collection   Eight class sets of video-stimulated recall interviews were conducted with teachers and three representative class students. (Note: due to time constraints experienced in one set of class interviews, two students only were interviewed). These students were chosen collaboratively by the teacher and researcher. Each student was selected because of their willingness to participate and the teacher's perception that they would provide thoughtful responses during the interviews. The interviews were designed to complement the video data through a semi-structured approach (Fontana & Frey, 1994). This approach was used to diffuse the possibility of extraneous explanations for events and to focus interviewees on particular behaviour episodes within the video data while still allowing for respondent elucidation. Video-stimulated recall (Clarke, 1998; Pirie, 1996; Meade & McMeniman, 1992) was used to trigger respondents' reliving and reflections of particular sequences of events.

Teacher and student interviews were held as soon as practicable after the video-recorded lesson (Meade & McMeniman, 1992; Morine-Dershimer, 1983; Nespor, 1984a). Before conducting each interview, the researcher provided a standard introduction to the video-stimulated recall interview process for both teachers and students. The interviews were structured similarly to Nespor's (1984a) techniques and in recognition of Calderhead's (1981) views that despite some reservations the technique has the potential to "explore more fully the cognitive aspects of classroom behaviour, and perhaps to develop new insights into the nature of teaching" (216).

The teacher interviewees were provided with the introductory dialogue and made aware that the interview process was to be informal. The informal nature of the interview process was considered important in recognition of the unusual nature of the task imposed upon the teacher interviewee (Nespor, 1984b). That is, the teachers were asked to view the video recording of their class and to comment on any teaching strategies they identified themselves as using. The teacher was instructed to stop the playback at this point and to respond to three questions:
– How do you describe the strategy?
– Why you decided to use the strategy?
– What effect do you believe the strategy has on student learning?

Strategy was used as a descriptor for teacher actions rather than tactics, in recognition of the widespread acceptance of strategy as a term for describing such teaching practices (Jonassen, Grabinger & Harris, 1990). The researcher asked the teacher for further information in order to clarify teacher explanations of the strategies identified. In addition, teachers were asked to advise the researcher of differences noted between their reasons for using particular strategies during the lesson and their reflective interpretations of strategy use as they observed the video recording. Teachers were encouraged to comment on any aspect of the video footage of the lesson that they felt was of interest to them. This was suggested in recognition of the novelty of the video-stimulated recall interview environment

(Nespor, 1984b). When the teacher did not choose to stop the footage at classroom situations deemed to be of interest to the researcher, the researcher adopted a proactive role, stopping the footage and asking for teacher comment. The researcher noted these points of interest within an observation sheet during filming of the technology lesson. Teacher interviews were concluded with the teacher being asked to provide details of their teaching experiences and an account of how they interpreted their teaching role within the videotaped lesson.

The time taken for teacher interviews ranged between one and two hours. The duration depended upon the length of the videotaped lesson and the extent of teacher comment. Several areas of repetitive classroom video footage, in terms of teacher and student activity (e.g. the teacher monitoring students as they engaged with their design and technology projects), was disregarded because of limitations in the amount of available teacher time.

Student interviews were conducted informally, though with more researcher direction than were teacher interviews. Students were asked to comment on specific instances within the recorded footage identified by the researcher. The researcher had previously identified these instances (i.e. the time within the footage) on an observation form. The time slip function of the DVD player facilitated movement through the footage from identified point to point with minimal time wastage. At these particular points in the lesson video recording, students were asked to explain:
- What they were thinking about during or immediately after that point in the lesson?
- What they did as a result of teacher actions at that point in the lesson?
- What they thought the teacher was trying to achieve at that point in the lesson?

In each case, if further clarification was required the students were asked to elaborate on their responses. As with teachers, students were encouraged to extend their comments beyond the above questions if they wished to do so. Student interviews ran for approximately fifteen minutes each. This allowed for adequate setup and introduction time for each of the students during an interview set (i.e. one videotaped lesson).

The interviews were audio recorded for later transcription. The audio recorder captured all sound (i.e. interviewee and interviewer comments, questions and responses and video footage audio) for the duration of the interview. This negated the necessity of logging teacher comments to particular times within the video footage. The links provided through the interview audio recordings, allowed the researcher to correlate the teachers' responses with the video for later interview transcriptions using Transana (The Board of Regents of the University of Wisconsin System, 1996-2003). Transana is a software program designed to enable the transcription and qualitative analysis of video and audio data. Similarly, the The Transana program allowed each series of student responses to be linked to the same video as was the teacher's responses.

Video-Stimulated Recall Interview Data Analysis The study followed the interview data-analysis technique used in Moallem's (1998) study of an expert teacher's thinking and instructional design. Moallem allowed categories to emerge from the data and sought to define the links between categories and subcategories in an effort to discover "core categories which became guides to further data collection and analysis" (43). In addition, codes were compared between different sources of data in order to seek their continual refinement. This process produced a representative set of themes that bound together the different components of Moallem's (1998) study.

The audio recordings were transcribed into Transana (The Board of Regents of the University of Wisconsin System, 1995-2003). During transcription, teacher and student dialogue was linked to the video data via time codes. These time codes embed the transcript to link teacher and student comments relating to particular teaching strategies (tactics), with the corresponding episode within the video recording.

Transana incorporates a facility for key-word searches of the transcribed data and can isolate sections of the video into meaningful groups based on these key words. The key words can be increasingly refined to allow for the generation of units of meaning as they evolve from an interpretation of the interview data (Cohen, Manion & Morrison, 2000; Moallem, 1998).

These key-words form codes (categories) that reduced the data into meaningful portions, which were used to analyse similar episodes from within alternate interview transcriptions (Moallem, 1998). Redundant codes were dropped as the raw data became incorporated within other more representative codes formulated specifically for the purpose. The aim of this process was to refine codes and sub-codes until accurate descriptions of technology education teaching and learning activities were evident between the various sets of interviews (Cohen, et al, 2000). This process is defined as being a system of constant comparison (Strauss, 1987 in Moallem, 1998; LeCompte & Preissle, 1993 in Cohen, et al, 2000) and is used to identify, compare and refine data categories as they emerge from the data (Moallem, 1998). In this study, the code categories are linked to activity system components (Jonassen, 2000) in order to frame the generated codes within a series of systemic models of technology classroom interrelated activity.

An analysis of the data (transcriptions) provided through the eight video-stimulated recall interviews was investigated for similar themes. The methodology, as described previously, revealed eleven themes emerging from within the data. These themes are indicative of each teacher's interpretation of the activities that emerge from within their respective technology lesson. The themes are provided in Table 1 and are listed under headings that segregate each theme into an activity theory component (i.e. tools, subjects, rules, or division of labour) (Jonassen, 2000). In addition, each theme carries with it an inherent purpose and it is the sum of these purposes that facilitate an interpretation of the teacher's tactics within each lesson. Student interviews allowed the researcher to confirm the teacher's interpretation of the intended learning outcomes from the student's perspective.

Table 1. Video-stimulated interview themes

TOOLS
FT- Focusing techniques (how the teacher focuses students to engage with their learning activities)
TM- Teacher Modeling (problem-solving, skills, attitudes etc.)
TLRW- Teacher linking the project or activity with the real world (reflected in the goals of the activity i.e. vocational skills development, thinking skills or conceptual development)
TF- Teacher facilitator (monitoring, cueing, hinting, providing answers – showing or telling)
SUBJECTS
SLA- Student learning attitudes (linked to student ability, motivation, and enjoyment)
ISLN- Individual student's learning needs (knowing the students and how they learn best, building rapport with the students)
RULES
TBASL- Teacher's beliefs about students and learning
TP- Teacher's philosophy about the purpose of the learning activity (why should students be involved in this activity, what is the ultimate goal, purpose or outcome of the activity)
CD- Classroom discipline (keeping students on task, how is class discipline maintained)
DIVISION OF LABOUR
R- Responsibility for learning (is the teacher or the student responsible for the learning outcomes)
TSI- Teacher student interaction (the teacher interacting with individuals, with groups of students or with the whole class)

Additionally, the researcher summarised each set of video-stimulated recall interviews (i.e. one teacher and three students) for two reasons. Firstly, to promote an image in the researcher's mind of the rationale behind each teacher's technology lesson and the teaching tactics and philosophy employed in the facilitation of lesson activities (Cohen, Manion & Morrison, 2000). Secondly, to provide phase 3 teachers with a condensed format of evidence relating to teaching tactics and philosophies that appear to impact upon the press of the technology learning environment for an increase in students' use of higher-order thinking.

Research results: Video and video-stimulated recall interviews

Technology education classes that provided data for phase 2 of the study varied significantly in respect to the degree that students within each perceived the press for FOCHP and SOCHP. This enabled the researcher to correlate evidence collected from both video and video-stimulated recall interviews with the press of the learning environment for particular types of student thinking. The results of this analysis provided support for a model of learning and teaching in technology education that appeared to support student higher-order thinking.

This chapter highlights the analysis of video-stimulated recall interview data gathered from within two of the six technology education classes investigated. Class 91's results for SOCHP and FOCHP were indicative of a technology learning environment supportive of student higher-order thinking (i.e. high for SOCHP and low for FOCHP). Class 12's results for SOCHP and FOCHP were indicative of a technology learning environment that was not supportive of student higher-order thinking (i.e. high for FOCHP and low for SOCHP).

Class 91 Video-Stimulated Recall Interviews Analysis The video-stimulated recall interview data analysis resulted in the allocation of teacher responses into the eleven code categories (See Table 1). These results were then summarised to provide a description of the teacher's tactics and philosophy used during the lesson.
- The teacher believes that each student is central in regards to their own learning activity.
- The teacher focuses students by questioning and probing for responses that clarify students' own understanding of the underlying principles of the activity.
- The teacher may not provide the correct answer for students but rather cues students to consider the possibilities and to make decisions based on their own understanding and research.
- The teacher makes judgements in regard to the level of support he provides to individual students. Success with too much support is a poor second place to failure and subsequent student reflection that leads to an understanding of why failure has occurred.
- The students are made aware that failure is a positive learning experience and as long as reflection and understanding results, students can still gain good results (assessment).
- The teacher models problem-solving for the students. He verbalises his thought processes whenever he has cause to seek clarification from the students (i.e. difficult problems are sorted out in front of the students in an explicit manner).
- The teacher links student activity with real world situations wherever possible.
- Students are encouraged to explain / present their knowledge to other students (i.e. the students become the experts).
- Students work generally by themselves on their own design projects.
- The teacher shares the responsibility with the student, though the student accepts that regardless of success or failure the learning is up to them.
- Students are expected to take responsibility for their own learning.
- The students understand the benefits of the teacher's tactics in regards to theirs and others learning.

Video-stimulated recall interviews with students from this class revealed that the students are well attuned to the teacher's intentions in this lesson. Generally the students feel comfortable and enjoy the atmosphere of the class. The teacher places the bulk of responsibility for learning with the students, although the students understand that the teacher is there to provide support for them in situations where their knowledge and understanding proves to be inadequate.

Class 12 Video-Stimulated Recall Interviews Analysis The video-stimulated recall interview data analysis resulted in the researcher allocating the teacher's responses into the eleven code categories. These results supported a summary of the teacher's tactics and philosophy used during the lesson.
- The teacher believes that each student should be encouraged to do good quality work and that it is his role to set the standard.
- The teacher focuses students by providing advice and demonstrating processes for students.
- The teacher provides the correct answer for students and seeks their global understanding of the processes involved in construction rather than demonstrating processes step by step for students.
- The teacher provides examples for students to facilitate their learning activity.
- The teacher links student activity with real world knowledge of things related to their projects.
- The teacher makes judgements in regard to the level of quality he will accept in relation to each student's project.
- The students are made aware that they are to ask for advice and to seek approval for stages of their project so mistakes are kept to a minimum. This activity is linked to students' motivation for good grades (i.e. students who don't do this are generally regarded as not caring about good grades).
- The teacher monitors student learning activity so that he can advise students as to the correct construction methods. If a mistake is made consistently he decides when to instruct the whole class in order to correct the error.
- The teacher links student activity with his own experiences of what works best for students in similar situations.
- Students are encouraged to ask (generally the teacher) for assistance in situations where their knowledge is inadequate.
- Students work generally by themselves on their own design projects.
- The teacher shares the responsibility with the student though the student considers that the teacher is the judge of quality and success or failure is up to the teacher.
- The students understand the benefits of listening to what the teacher tells them about how to do things.
- Students expect that the teacher will solve their problems (i.e. if you don't know ask the teacher).

Student video-stimulated recall interviews revealed that the students enjoy the subject and they feel that the atmosphere of the class is positive. The students see the projects as being the motivating feature of the subject and in particular the surfboard etc (the project being constructed at the time of video recording) is a project that they have enjoyed making. The interviewed students enjoy working with their hands. One student linked this aspect of the subject with his desire to become a builder in later life. The students confirmed the teacher's use of real-world aspects of the project to support their learning. The students felt that the teacher should be consulted when they have problems or encounter difficulties. Sometimes they would ask their friends, although confirmation is usually sought from the teacher also. The students interpret the teacher's activity of providing examples and judgements of quality as being important aspects of the lesson. The students' impressions of the purposes for the teacher's actions and activities were representative of the teacher's own interpretations.

CONCLUSION

The aforementioned analysis of video and video-stimulated recall interview data for classes 12 and 91 are indicative of the variation in press of the learning environments for different types of student thinking. It is apparent that each teacher (class 91 and class 12) acknowledges different degrees of responsibility for student learning outcomes. Or perhaps more accurately, each teacher creates or facilitates a learning environment that differs in terms of the extent that the environment presses students to acknowledge their own responsibility for learning outcomes. In both cases teachers support students, it is the focus of this support that dictates the response of students regarding the types of thinking used.

Video-stimulated recall interview techniques provides learning environment researchers with an approach to data collection and analysis that is intensive and authentic. Authentic in the sense that the video footage and the stimulated recall interviews are structured in response to what actually happens in classrooms. However, a camera's eye view of classroom activity is an artificial snapshot of the lesson and should not be argued to be a replacement for experiencing the lesson first hand. Additionally, a researcher is not of the environment. A researcher is an outsider who seeks to interpret the learning environment by looking inwardly at the activities of the teacher and students. The value of video-stimulated recall is that an interpretation of the lesson's activities (albeit from a certain theoretical perspective) is negotiated through evidence that is generated from within the environment and from the actual inhabitants' experiences as they naturally occur from within the setting. The video footage stimulates informed dialogue in respect to what happened during the lesson and why.

This chapter has provided an overview of a research study that integrates qualitative and quantitative methods of investigation through a number of interconnected and mutually supported phases. The research design addresses the complexities of technology education learning environments through the combination of these methods and phases.

The study is limited by factors that constrain all educational research, for example, institutional constraints (e.g. researcher access, time limits, etc), resource constraints (e.g. limited equipment, staff, money, etc) and ethical constraints (e.g. difficulties with gaining informed consent, etc) (Krathwohl, 1993). However, within these limits this research project has maintained an approach, which identified and reduces the threats to both internal and external validity. In addition, the study phases were mutually supportive and interconnected in such a way as to promote the strengths and limit the weaknesses of both the quantitative and the qualitative methods used in this inquiry.

Video-stimulated recall interviews provide a rich source of qualitative data and along with the video data, support a judicious understanding of the learning environments under investigation. This understanding coupled with quantitative evidence of the press of the technology learning environment for different types of student thinking, supports practically based proposals for the implementation of appropriate teaching tactics that promote student higher-order thinking in technology education classrooms.

REFERENCES

Babbie, E. (2001). *The practice of social research (9th Edition)*. Wadsworth Thomson Learning.

Bernard, R. H. (2000). *Social research methods: Qualitative and quantitative approaches*. Sage Publications, Inc.

Calderhead, J. (1981). Stimulated recall: A method for research on teaching. *British Journal of Educational Psychology, 51*, 211-217.

Campbell, D. T., & Stanley J. C. (1963). *Experimental and quasi-experimental designs for research*. Rand McNally College Publishing Company Chicago.

Clarke, D. (1998). *The classroom learning project: Its aims and methodology*. Paper presented at the 1998 conference of the Australian Association for Research in Education, November 30, 1998. Retrieved from the World Wide Web on 10/9/2002. URL: http://www.aare.edu.au/98pap/cla98053.htm

Cohen, L., Manion L., & Morrison K. (2000). *Research methods in education (5th Edition)*. Routledge Falmer, London and New York.

DeMiranda, M. A., & Folkestad J. E. (2000). Linking cognitive science theory and technology education practice: A powerful connection not fully realized. *Journal of Industrial Teacher Education, 37*(4). Retrieved from the World Wide Web on 10/5/2001. URL: http://scholar.lib.vt.edu/ejournals/JITE/v37n4/demiranda.html

Elshout-Mohr, M., Van Hout-Wolters, B. & Broekkamp, H. (1999). Mapping situations in classroom and research: eight types of instructional learning episodes. *Learning and Instruction, 9*, 57-75.

Eggleston, J. (1992). *Teaching design and technology:* Open University Press.

Fontana, A., & Frey, J. H. (1994). Interviewing: The art of science. In N. K. Denzin & Y. S. Lincoln (Ed), *Handbook of qualitative research* (pp. 361-376). Sage Publications, Inc.

Fritz, A. (1996). Reflective Practice: Enhancing the outcomes of technology learning experiences. *The Journal of Design and Technology Education, 1*(3), 212-217.

Gump, P. V. (1991). School and classroom environments. In D. Stokols & I. Altman (Ed). *Handbook of environmental psychology* (pp. 691-732). Krieger Publishing Company, Malabar, Florida.

Gump, P. V. (1978). Big schools, small schools. In R. G. Barker and Associates, *Habitats, environments, and human behavior* (pp. 245-256). Jossey-Bass Publishers.

Hansen, R. E. (1995). Five principles for guiding curriculum development practice: The case of technological teacher education. *Journal of Industrial Teacher Education, 32*(2), 30-50.

Herschbach, D. R. (1998). Reconstructing technical education. *Journal of Industrial Teacher Education, 36*(1). Retrieved from the World Wide Web on 10/5/2001. URL: http://scholar.lib.vt.edu/ejournals/JITE/v36n1/herschbach.html

Herschbach, D. R. (1995). Technology as knowledge: Implications for instruction. *Journal of Technology Education, 7*(1), 31-42.

ITEA, International Technology Education Association (2000). *Standards for technological literacy: Content for the study of technology, executive summary*. Retrieved from the World Wide Web on 13/08/02. http://www.iteawww.org/TAA/PDF/Execsum.pdf

ITEA, International Technology education Association (1996). *Technology for all Americans: A rationale and structure for the study of technology*. Retrieved from the World Wide Web on 13/08/02. http://www.iteawww.org/TAA/PDF/Taa_RandS.pdf

Jonassen, D. H. (2000). Revisiting activity theory as a framework for designing student-centered learning environments. In D. H. Jonassen & S. M. Land (Ed), *Theoretical foundations of learning environments* (pp. 89-121). Lawrence Erlbaum Associates, Inc.

Jonassen, D. H., Grabinger, R. S., & Harris, N. D. (1990). Analysing and selecting instructional strategies and tactics. *Performance Improvement Quarterly, 3*(2), 29-47.

Johnson, S. D. (1996). Technology education as the focus of research. *The Technology Teacher*, May/June, 47-49.

Kemp, W. H. (Ed) & Schwaller, A. E. (Eds.) (1988). *Instructional strategies for technology education* (pp 143-165). Glencoe Publishing Company.

Krathwohl, D. R. (1993). *Methods of educational and social science research*. Longman Publishing Group.

Lauda, D. P. (1988). Technology education. In W. H. Kemp & A. E. Schwaller (Eds.), *Instructional strategies for technology education* (pp.3-15). Glencoe Publishing Company.

Lee, S. (1996). Problem-solving as intent and content of technology education. Paper presented at the International Technology Education Association 58th Annual Conference in Phoenix, Arizona, March 31-April 2, 1996. (Eric Document Reproduction Service No. ED 391 959).

Leshin, C. B., Pollock J., & Reigeluth C. M. (1992). *Instructional design strategies and tactics*. Englewood Cliffs, New Jersey: Educational Technology Publications, Inc.

Levin, J. R., & O'Donnell, A. M. (1999). What to do about educational research's credibility gaps? *Issues in Education, 5*(2), 177-229.

Meade, P., & McMeniman, M. (1992). Stimulated recall – An effective methodology for examining successful teaching in science. *Australian Educational Researcher, 19*(3), 1-11.

McCormick, R. (1997). Conceptual and procedural knowledge. *International Journal of Technology and Design Education, 7*, 141-159.

McCormick, R. (1996). Instructional methodology. In J. Williams and A. Williams (Ed), *Technology education for teachers*. (pp. 63-92). Macmillan Education Australia Pty. Ltd.

McCormick, R. & Davidson, M. (1996). Problem solving and the tyranny of product outcomes. *The Journal of Design and Technology Education, 1*(3), 230-241.

McCormick, R., Murphy, P. & Hennessy, S. (1994). Problem-solving processes in technology education: A pilot study. *International Journal of Technology and Design Education, 4*, 5-34.

Moallem, M. (1998). An experts teacher's thinking and teaching and instructional design models and principles: An ethnographic study. *Educational Technology Research and Development, 46*(2), 37-64.

Morine-Dershimer, G. (1983). *Tapping teacher thinking through triangulation of data sets*. R & D Report No. 8014. Texas University, Austin, Research and Development Centre for teacher Education. (Eric Document Reproduction Service No. 251 434).

Nespor, J. K. (1984a). *The teacher beliefs study: An interim report. Research on the social context of teaching and learning*. R & D Report No. 8020. Texas University, Austin, Research and Development Centre for teacher Education. (Eric Document Reproduction Service No. 270 446).

Nespor, J. K. (1984b). *Issues in the study of teachers' goals and intentions in the classroom*. R & D Report No. 8022. Texas University, Austin, Research and Development Centre for teacher Education. (Eric Document Reproduction Service No. 260 075).

Newberry, P. B. (2001). Technology education in the U.S.: A status report. *Technology Teacher, 61*(1), 8-12.

New Zealand Ministry of Education (1995). *Technology in the New Zealand Curriculum*. Retrieved from the World Wide Web on 6/05/2004. URL:
http://www.minedu.govt.nz/web/downloadable/dl3614_v1/tech-nzc.pdf

Perkins, D. (1993). Creating a culture of thinking. *Educational Leadership, 51*(5), 98-99.

Pirie, S. E. B. (1996). Classroom video-recording: When, why and how does it offer a valuable data source for qualitative research? Paper presented at the annual meeting of the North American Chapter of the International Group for the Psychology of Mathematics Education. Panama City, Fl, October 14, 1996. (Eric Document Reproduction Service No. 401 128).

QCA, Qualifications and Curriculum Authority (1999). Design and Technology: The National Curriculum for England www.nc.uk.net. Retrieved from the World Wide Web on 13/08/02. URL: http://www.nc.uk.net/download/bDT.pdf

QSA, Queensland Studies Authority (2002). *Industrial technology and design subject area syllabus.* Retrieved from the World Wide Web on 2/08/2002. URL:
http://www.qsa.qld.edu.au/yrs1_10/kla/other_studies/sas/itd_ed.pdf

QSA, Queensland Studies Authority (2003). *The technology years 1 to 10 syllabus.* Retrieved from the World Wide Web on 6/05/2004 URL:
http://qsa.qld.edu.au/yrs1_10/kla/technology/pdf/syllabus.pdf

Resnick, L. B. (1987). *Education and learning to think.* Washington, D.C.: National Academy Press.

Rothwell, W. J. & Kazanas, H. C. (1992). *Mastering the instructional design process: A systematic approach.* Jossey-Bass Publishers, San Francisco.

Sanders, M. (2001). New paradigm or old wine? The status of technology education practice in the United States. *Journal of Technology Education, 12*(2), 35-55.

Schoggen, P. & Schoggen, M. (1988). Student voluntary participation and school size. *Journal of Educational Research, 81*(5), 288-293.

Schultz, A. E. (2000). Cognitive psychology as the basis for technical instruction? I don't think so. *Journal of Industrial Teacher Education, 37*(4). Retrieved from the World Wide Web on 10/5/2001. URL: http://scholar.lib.vt.edu/ejournals/JITE/v37n4/schultz.html

Shield, G. (1996). Formative influences on technology education: The search for an effective compromise in curriculum innovation. *Journal of Technology Education, 8*(1). Retrieved from the World Wide Web on 10/8/2002. http://scholar.lib.vt.edu/ejournals/JTE/v8n1/Shield.html

Slavin, R. E. (1999). Educational research in the 21st Century: Lessons from the 20th. *Issues in Education, 5*(2), 261-266.

Stevenson, J. C. (1998). Performance of the cognitive holding power questionnaire in schools. *Learning and Instruction, 8*(5) pp. 393-410.

Stevenson, J. C. & Evans G. T. (1994). Conceptualization and measurement of cognitive holding power. *Journal of Educational Measurement, 31*(2) pp. 161-181.

Stevenson, J. C. & McKavanagh C. (1994). Development of student expertise in TAFE colleges. In J. Stevenson (Ed.), *Cognition at work. The development of vocational expertise* (pp. 169-197). National Centre for Vocational Education Research Ltd.

The Board of Regents of the University of Wisconsin System (1995-2003). *Transana Version 1.22.* Originally created by Fassnacht C., now developed and maintained by Woods D. K., Wisconsin Centre for Education Research, University of Wisconsin-Madison. © The Board of Regents of the University of Wisconsin System. Retrieved from the World Wide Web on 10/10/2002. Last accessed 2/07/2004. URL: http://www2.wcer.wisc.edu/Transana/Transana

Walmsley, B. D. (2004). Student decision-making: The teacher's role. Paper presented at The Pupils Attitudes towards Technology (PATT) 14 Conference, Albuquerque, New Mexico, USA, 18-19 March, 2004. Pupil's decision-making in technology: Research curriculum development and assessment. URL:http://www.iteawww.org/PATT14/Walmsley.doc

Walmsley, B. D. (2003). Partnership-centered learning: The case for pedagogic balance in technology education. *Journal of Technology Education, 14*(2), 56-69.

Walmsley, B. D. (2001). *Technology education learning environments and higher-order thinking.* Unpublished Honours Thesis. Griffith University, Brisbane, Queensland, Australia.

Wicker, A. W. (1984). *An introduction to ecological psychology.* Cambridge University Press.

Williams, J. (2000). Design: The only methodology of technology? *Journal of Technology Education, 11*(2) Spring 2000. Retrieved from the World Wide Web on 10/9/2000. URL: http://scholar.lib.vt.edu/ejournals/JTE/v11n2/williams.jte-v11n2.html

Wollman-Bonilla, J. E. (2002). Does anybody really care? Research and its impact on practice. *Research in the Teaching of English, 36*, Feb, 311-326.

Young-Hawkins, L. & Mouzes M. (1991). *Transforming facilities: Industrial arts to technology education.* Paper presented at the American Vocational Association Convention, Los Angeles, CA, December 7, 1991. (Eric Document Reproduction Service No. ED 341 868).
URL: http://jennifer.lis.curtin.edu.au/thesis/available/adt-WCU20020502.121823/

Zohar, A., Degani, A., & Vaaknin, E. (2001). Teachers' beliefs about low achieving students and higher order thinking. *Teaching and Teacher Education, 17 (4)*, 469-485.

Zuga, K. F. & Bjorkquist, D. C. (1989). The search for excellence in technology education. *Journal of Technology Education, 1*(1). Retrieved from the World Wide Web on 10/9/2000. http://scholar.lib.vt.edu/ejournals/JTE/v1n1/zugabjor.jte-v1n1.html

Bradley Walmsley
Faculty of Education
Griffith University
Australia

HOWARD MIDDLETON

EXAMINING DESIGN THINKING

Visual and Verbal Protocol Analysis

INTRODUCTION

This chapter examines the way in which design students and practicing architects engage in the act of designing. The approach taken is cognitive and the intention is to outline a research methodology that examines two related aspects of design activity. The first is the use designers make of procedural knowledge, often described as "knowledge how". The second is the ways in which designers employ visual mental images, in designing. Finally, methods for examining how these two aspects of design activity work together are explored.

The chapter draws on two studies of design thinking. The first study involved four high school technology education students, while the second involved three architects of varying levels of expertise. For the purposes of this chapter, however, one subject from the first study, and two subjects from the second study will be used. The reason for this approach is that it will provide a clearer illustration of the research approach while keeping within a reasonable size for the chapter. Moreover, the two studies were linked in that the subjects in both studies were given the same architectural design brief and in the analysis they were defined as being located on levels of the same novice-expert categorization developed by Dreyfus and Dreyfus (1986). The Dreyfus and Dreyfus categorisation comprises novice, advanced beginner, competent, proficient and expert. The school student was defined as an advanced beginner, and the two architects were categorised as competent and expert.

DESIGNING

Designing involves people in developing ideas that are represented by words on paper, or in conversations, and by images in their heads and on paper, or computer screens. The multi-modal nature of design activity, like much technological activity, presents challenges for researchers in education. The challenge occurs because educational research is based largely in the humanities or social sciences, where the tradition has developed that all data that is necessary for understanding human thinking and action can be represented adequately by words. Indeed, some theorists (Pylyshyn, 1981) have argued that information in memory is only stored in abstract, word-like representations. More recent research (Anderson, 2006; Kosslyn, 1994) has provided strong neurological evidence to support the multi-modal view of memory, however, the tradition of the pre-eminence of word-based evidence remains strong.

Design can be conceptualised as a particular form of problem-solving and Middleton (1998) has proposed a model to account for the particular characteristics of design problems. In the model, problems are conceptualised as occurring in a problem space, containing a problem zone, a search and construction space and a satisficing zone. The aim in problem-solving is taken to be to move from the problem zone, through the search and construction space to the satisficing zone. Design involves a series of complex interactions between many variables that make it inappropriate to use more quantitative methods of analysis. (Yin, 1994; Gay, 1993). As a consequence, the research described in this chapter constitutes a case study approach.

Selection of subjects

The aims of the studies were firstly, to examine the way people solve design problems and the use they made of visual mental images; and secondly, to examine the nature of both design expertise and the transition from novice to expert designer. The selection of subjects was based on five considerations. The first was the need to select subjects, over the two studies who could provide data on identifiable stages in the progression from novice to expert. However, in designing, a certain level of knowledge is required before it is possible to attempt, in a meaningful way, to solve a design problem. For example, knowledge of some of the conventions of designing, and knowledge of how to sketch is required. For this reason, novice subjects were not selected. That is, subjects who could be considered "one who is new to the circumstances in which he [sic] is placed; a beginner" (Onions, 1974, p. 1418).

To provide an adequate characterisation of the remaining stages between novice and expert, the categories devised by Dreyfus and Dreyfus (1986) were employed. The stages in the Dreyfus and Dreyfus classification are: novice, advanced beginner, competent, proficient and expert. For Study I, the design student was regarded as an advanced beginner, and the two architects from Study II were categorised in turn as competent, and expert.

The second consideration was the need to select subjects for whom the design problem represented a close approximation to the kind of problem they would be expected to solve during their normal work or learning experience. That is, it needed to be seen as authentic by the subjects. The third consideration was the need to select a design field where the prospective subjects were either accustomed to, or comfortable with, 'thinking aloud' when solving design problems in front of an observer. (This consideration determined that, in the case of the high school student, selection was voluntary and, in the case of the architects, that all subjects had experience of domestic architecture, where 'thinking aloud' in front of a client is normal practice). The fourth consideration was the need to select, for Study II, subjects who covered a range in terms of design experience. Thus, subjects in Study II included architects with experience ranging from one year to thirty eight years. The last consideration was the need to select subjects whose verbal ability was sufficient to ensure that the data collected were not limited by gaps in verbal ability. The selection of professional architects and students who had successfully completed atleast eight years of schooling was considered to have addressed this issue.

Data

Two forms of data were collected for the research. The first was verbal think-aloud protocols (Ericsson & Simon, 1993). These are defined as any thoughts that emerge in the designer's mind and are verbalized as they engage in designing. The second type of data were video images of the design being created on paper and any associated hand movements.

Selection of settings

The problem-solving data were collected in a variety of settings. The main criterion was that the setting be a natural one in the sense that it was a setting where such design activity regularly occurred. The settings and timing for the architects were chosen by the architects (one in his home studios, one in her office). The student protocols were collected during time scheduled for Technology education classes. The student protocols were collected in rooms adjacent to the classroom.

The design problem

The design problem needed to meet a number of specifications. Firstly, it had to be one that all subjects saw as authentic. To achieve this it was devised with the collaboration of an architect who was not a subject of the study, and with a high school, technology teacher. Secondly, it had to contain features that would make it complex for all participants, to avoid subjects using automated processes that tend not to be verbalised. Lastly, the problem had to be one that could accommodate school timetables, and the amount of time professional designers would be prepared to devote to such an activity. The design problem was as follows:

> An elderly female client, with limited funds, requires a detached dwelling on a 450m^2 urban Brisbane block. The site measures 15m X 30m with the shortest alignment fronting the street. The site falls away from the street at approximately 30^0, towards the South and is free of established trees. The best views lie to the South, due to the elevated nature of the site. The client suffers from an arthritic condition and has great difficulty negotiating stairs. She values her privacy but would like the house to have a light and airy atmosphere. She also wishes to take best advantage of natural light, breezes and passive solar energy. She is open to suggestions on the general form and

character of the house, and despite her age, could not be considered conservative.

Collecting the data

Each subject was informed that they would be required to solve an architectural design problem, and that they would be required to verbalise all and any thoughts that came into their head while they solved the problem. They were also asked to solve the problem in the normal way they would solve it as a class exercise, in the case of the students, or in professional practice, in the case of the architects. Subjects were also told that a video camera would record everything they said, and would record any sketches they made while solving the problem. They were informed that if they did not say anything for more than ten seconds, they would be asked: What are you thinking now? (Ericsson & Simon, 1993) Each subject was then handed a printed copy of the design brief, after which the video-recording commenced. Subjects were not told that the problem might be difficult, nor was any mention made of any need to be creative or to use visual mental images when solving the problem. At the conclusion of the problem-solving episode, sketches produced by the subjects were collected for later analysis, along with the videotapes.

It is worth noting that when the videotaping was trialled, the camera was located in front of the subjects, in order to collect details of non-verbal behaviour, such as facial movements, as well as the generation of the design sketch and verbalisations. This proved to be quite intimidating for the subjects and the position of the camera was changed to one where it was located behind the subject and to the left or right. This captured the emerging design, hand movements and voice, but not all other non-verbal data (eg facial expressions). It did, however, allow the subjects to relax and forget they were being recorded.

Analysing the data

Data from the two studies were analysed to establish firstly, the cognitive procedures employed in solving complex problems and secondly, to identify both the characteristics of mental images used in solving design problems, and the function they served in the problem-solving process. In addition, differences across problem-solvers of different degrees of expertise were examined. The following section presents the framework used for coding the verbal protocols which allowed conclusions to be drawn, based on the model of problem-solving described earlier in this chapter.

Preparation of data for analysis

The data source consisted of a videotape recording of the problem-solving activity of each subject which included a continuous recording of sketching activity and all verbalisations for each of the subjects in Studies I and II. The verbal records of problem-solving were transcribed and segmented on the basis of achieving the smallest unit of meaning that constituted an instance of a general process, which often meant segmenting at the level of clauses. Cues for segmenting protocols included pauses and intonation, as well as syntactical cues of complete phrases and sentences (Ericsson & Simon, 1993). Thus, the segmentation process followed recognised procedures.

Coding the verbal protocols for cognitive procedures

Each segment of the verbal protocols of problem-solving activity was coded in terms of cognitive procedures into one of ten types of procedures, which was then located within one of three larger categories. The relationship between the procedures and categories of procedures is discussed in Middleton (1998). This relationship is summarised in Table 1, and explained more fully in the following section.

Table 1 Categories of procedures and procedures

Category of procedure	Generation	Exploration	Executive control
Procedure	Retrieval Synthesis Transformation	Exploring constraints Exploring attributes	Goal setting Strategy formulation Goal switching Monitoring Evaluation

Executive control

The category of procedures called executive control contains the procedures of goal setting, which is defined as the process of establishing an overall goal (Anderson, 1993) strategy formulation, which is defined as an indication of a general heuristic for approaching a problem or parts of a problem (Schon, 1990) goal switching, which is defined as a change of attention from one aspect of the problem to another (Eckersley, 1988), monitoring which is defined as the process of checking the progress of problem-solving to establish if goals are being achieved and solution constraints satisfied (Chan, 1990; Simon, 1975), and evaluation which is defined as normative statements about proposals, attributes or strategies (Finke, Ward & Smith, 1992).

Generation

The generation category refers to the procedures of <u>Retrieval</u>, which is defined as retrieval of knowledge from long-term memory (Finke, 1989), or the retrieval of information from visual perception of physical objects or visual displays (Larkin & Simon, 1987), <u>synthesis</u>, which is defined as the formulation and articulation of a specific proposal to solve a problem or sub-problem, and <u>transformation</u>, which is defined as modifying a proposed idea to enable it to solve a particular problem (Weber, Moder & Solie, 1990).

Exploration

The exploration category refers to the procedures of <u>exploring constraints</u>, which is defined as the identification of aspects of the problem context or design proposals as containing elements that are perceived as adding to the complexity of the problem. (Described also as acknowledging constraints (Finke, Ward & Smith, 1992; Gross & Fleisher, 1984; Simon, 1970) and <u>exploring attributes</u>, which is defined as aspects of the problem context or proposed solutions that either facilitate problem resolution or define problem context (Finke, Ward & Smith, 1992).

167 it may protrude back again, into the line of the garage	[TR]
168 and can get a window here to the North	[SY]
169 Which will give us some sun in winter	[EA]
170 and that's about three metres	[SY]
171 that's quite, quite a decent sized bedroom	[EV]
172 the en-suite may also be accessed from this level as well	[EA]
173 I don't know if she will like that	[EA]
174 but if she is entertaining on this area people will need to use that bathroom	[GS]
175 so it may be a two-way bathroom	[SY]
176 and the stairs .... to go downstairs could fit in this area, in here	[SY]
177 should be fairly narrow	[SY]
178 so as not to take up too much of the room within the building	[EA]
179 we want to get as much of the rooms facing that way	[SY]
180 without the stairs coming into that area	[EA]

Note: Some of the segments in Figure 1 may appear to be incorrectly coded. For example, on the surface, segment 179: we want to get as much of the rooms facing that way, would appear to satisfy the requirements to be coded as goal setting [GS]. However, reference to the videorecording of the segment indicates that rather than espousing a general goal as the verbalisation suggests, the problem-solver was in fact creating the specific instantiation of the goal. That is, synthesising a specific solution element. In this instance, the subject was sketching the rooms referred to. Cross-referencing of data sources was used in this way to improve validity of codings.

Figure 1 Illustration of coding for cognitive procedures

Tentiles

After coding the segments, each problem-solving episode was divided into tentiles. That is, they were divided into ten equal portions. This is a common way of preparing verbal protocol data, however, it is common practice to divide on the basis of numbers of segments, for example, if there were one hundred segments, each tentile would contain ten segments. The presumption in such a practice is that meaning is only contained in words. However, for designing it was regarded as important to be able to capture pauses and changes in the rate of problem-solving. As a consequence, the tentile divisions were based on time, not number of segments.

It is important to note that the length or duration of protocols varied across subjects, and thus, tentiles were not equal in length or duration across subjects. However, each tentile covers the same relative proportion of problem-solving activity. Thus, tentile 10 covers the last ten percent of the problem-solving activity for each subject.

After dividing protocols into tentiles, the ten sub-categories were consolidated into the three major categories of design activity. The consolidation was carried out to provide an alignment with the geneplore categories developed by Finke, Ward and Smith (1995). These categories of cognitive procedures were accepted as signifying significant aspects of design cognition. In addition, the consolidation made possible a more general level of analysis and also facilitated the identification of trends, for example, trends in the movement from novice to expert.

Using these three categories, a spreadsheet was created with the vertical columns representing the numbers of procedures in each of the three categories of cognitive procedures and the horizontal rows representing the tentiles. From the spreadsheet, scatterplots were generated to facilitate analysis of the problem-solving process and comparisons between subjects. To illustrate, the scatterplot for subject A1 (Competent) is shown in Figure 2 and for A3 (expert) in Figure 3.

The scatterplots were analysed to establish where, in a problem-solving activity, individual subjects used particular categories of procedures and how these patterns of usage varied between subjects. The representation of cognitive procedures in the scatterplot is useful for analysis. For example, if we look at Figure 2, we can see problem-solving behaviours typical of a novice in a domain. That is, subject A1 engaged in rapid generation of solutions early in the problem-solving, without much exploration, or control processes. This quickly became problematic as things started to go wrong and he realised he needed to explore the issue more before proceeding.

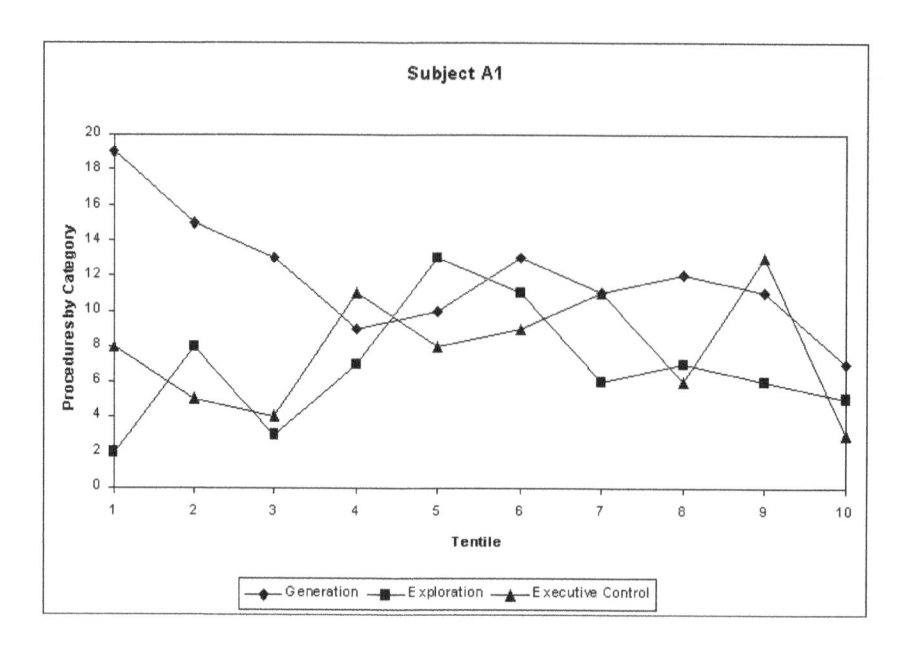

Figure 2 Scatterplot for Subject A1

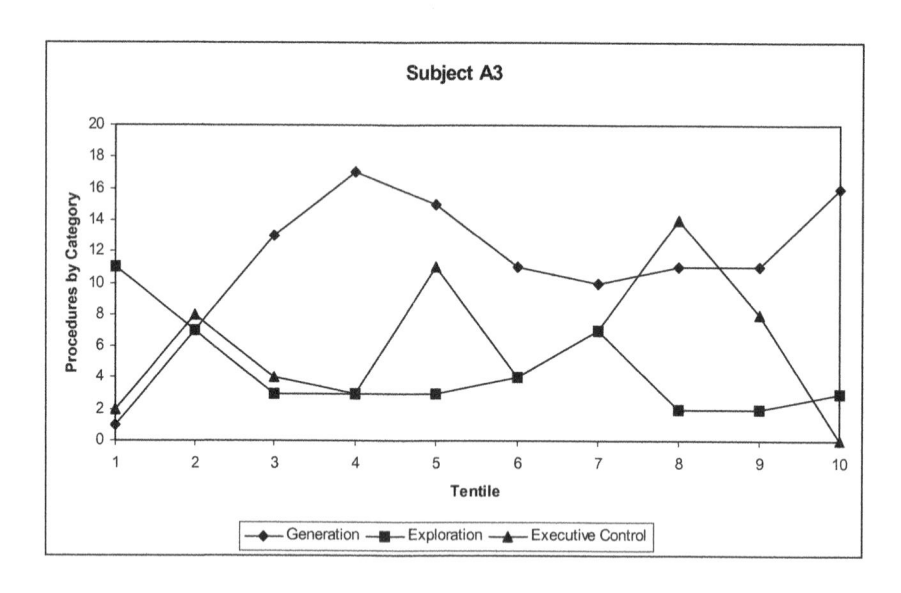

Figure 3 Scatterplot for Subject A3

Figure 3 illustrates very different problem-solving to that in Figure 2. It can be seen in Figure 3 that the expert engaged in significant exploration before engaging in problem-solving. However, once problem-solving commenced, progress was rapid, and the problem was largely resolved by tentile 4. This is typical for an expert in a domain. As can be seen in Figures 2 and 3, converting the design performance into tentiles and scatterplots is a very good way of making sense of verbal data.

After all think-aloud protocols were coded and analysed in terms of categories of cognitive procedures, the visual think-aloud protocols were coded and analysed in terms of cognitive representations, and this is covered in the following sections.

Coding the visual protocols for cognitive representations

Visual mental images, as a form of internal cognitive representation, constitute a phenomenon that is essentially unobservable in any direct sense (Shepard, 1978). However, it is argued that, on the basis of research and theorising (summarised in Finke, 1989), visual mental images share essentially the same internal mechanisms as images that are the product of perception (with a number of non-threatening exceptions). The equivalence between imagery and perception includes spatial, structural, functional and transformational features.

On the basis that mental images are equivalent to perceived images, it is then argued that sketches on paper, produced in response to a design problem, and in the absence of any objects or drawings to perceive and copy, represent the externalisation of visual mental images. Sketches were therefore regarded in this study as the primary evidence for the use of visual mental imagery in problem-solving in design.

The purpose of the imagery coding was to provide the basis for establishing when thinking was in a visual mode, the properties of different images, the function that the images performed in problem-solving, the variation in imagery types and imagery usage between different problem-solvers, and the instances of parallel processing, that is, instances when sketching suggested imaginal thinking while the concurrent verbalisation suggested verbal processing was also occurring.

Representation of the visual data

The coding for visual data involved taking the coded verbal data illustrated in Figure 1, and creating two columns where the verbal data are located in a left hand column with the contiguously generated sketches located on the right-hand column, as illustrated in Figure 4. To provide for a more fine-grained analysis, those sections of text that represented verbalisations that were produced while sketching was actually taking place were shown in italics. That is, while the horizontal lines defined the extent of the verbalisations generated concurrently with a particular sketch, they were unable to indicate pauses in sketching; italics were used to overcome this problem. The italicised text was used to determine if sketching was occurring with particular verbal segments. In instances where uncertainty occurred in terms of the coding of a particular verbal segment, the presence or absence of italics indicated whether the videotape record could provide any additional material to clarify coding.

```
027  now taking best advantage of the site ......
028  um, just looking at it as a block diagram
029  I'd have most living spaces and bedrooms on the northern side, so
030  I'll just indicate that's North for now
031  um ...... now with the best views ......
032  falls away from the street towards the South
033  the best views towards the South
034  difficult, difficult, difficult ................
035  I'll probably still keep the bedrooms on the northern side
036  but make it a very linear plan
037  so we take advantage of the views
038  as well as the northern aspect, um
039  which will also help with cross ventilation
040  so if we could have the carport at one end
041  and then have the lounge and kitchen, um
042  I need to know a bit more about the
      ..... client
043  but anyway
044  and then maybe we will spread
045  the ah bedrooms out along one edge
046  there's the bathrooms up
047  bathrooms and en-suite occupy ... um ......
048  possibly the eastern end of the building
049  so that's East
050  ok, so ............. let's see              Drawing 1
```

carport	lounge	
	kitche	E

Note: verbalisations that occurred concurrently with sketching are shown in *italics*

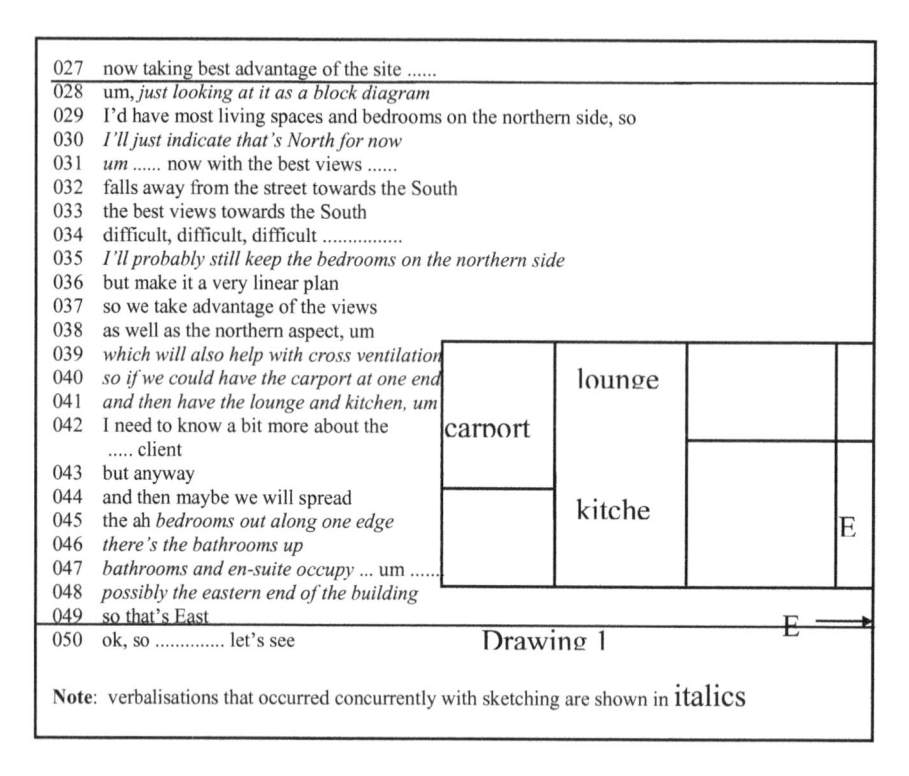

Figure 4 Illustration of initial representation of problem-solving data for cognitive representations

After the initial representation of the verbal and visual data, the protocols were divided into tentiles (as described above), and the sketches located on a horizontal tentile grid (illustrated in Figure 5). Three tentile grids were used, with one displaying the sketches and sketch descriptors for the student in Study I, and the other two displaying the sketches for the two architects in Study II. Location on the tentile grid allowed comparisons to be made across problem-solvers in terms of the location and function of particular types of sketches and the images from which they were produced. It also enabled links to be made with verbal data.

EXAMINING DESIGN THINKING

Analysis of visual protocols

The purpose of the analysis and, thus, of the coding of visual representations, was to establish what kinds of images were being used, their properties, what function they fulfilled in the problem-solving process and what relationship they had to procedures (identified through verbal data). The images were described in terms of four categories; sketch type, function, image type and properties. These are outlined below:

Sketch type - *provides a description of the final, static instantiation of an image. It is described in terms of architectural type (eg block plan, floor plan) and the presence of abstract labels.*

Function - *provides a description of the inferred role that the sketch played in the problem-solving performance.*

Image type - *provides a description of each image in terms of whether the image is static or dynamic, detailed or simple and two or three-dimensional.*

Properties - *provides description of additional features such as if an image is the result of a zooming process. An illustration of the representation of the visual data on the tentile grid for the student can be seen in Figure 5.*

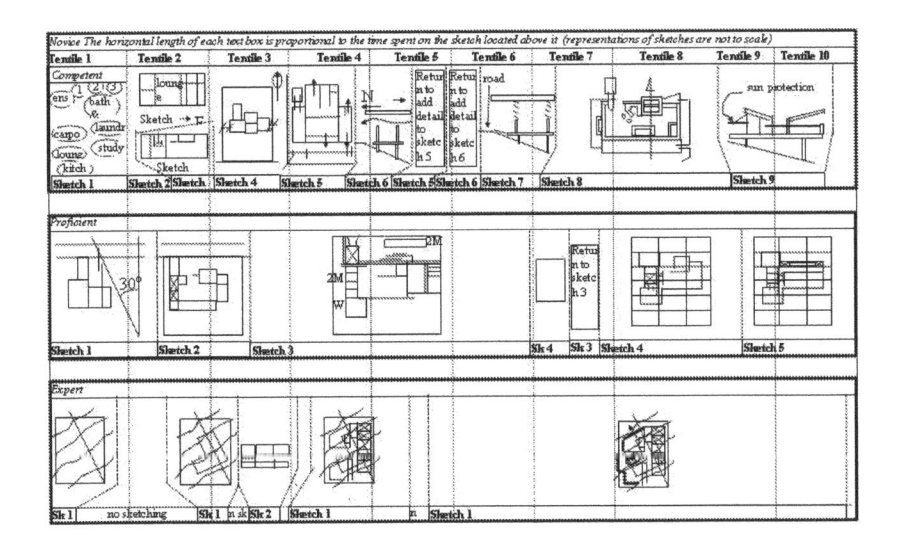

Figure 5. Visual Protocols

An examination of the visual data generated by the three subjects in both studies suggests that firstly, visual mental images facilitate designing by designers at all levels of expertise, and secondly, that imagery usage varies with expertise, with the expert, on the one hand, having access to rich and detailed images, that may be dynamic. The advanced beginner, on the other hand, having access to less detailed,

203

and generally static images. Thus, the studies suggest that the ability to produce visual mental images is an important design skill, regardless of the level of expertise and that development of design expertise is dependent on development of the ability to develop visual mental images.

Analysing the relations across verbal and visual data

To examine the relations between the verbal and visual representations of design problem-solving activity, two kinds of analyses were undertaken. Firstly, the data were examined in terms of the categories of procedures that occurred during image initiation. That is, the analysis examined the point at which each image commenced and related each of these instances to the cognitive procedure the problem-solver was deploying at the time. The reason for undertaking this analysis was the argument that under conditions of cognitive difficulty, problem-solvers will use the most efficient representation of a problem (Kaufmann, 1990; Sheikh, 1985). If the verbalisation associated with the commencement of a new sketch is an executive control procedure, this suggests both that the problem-solver is experiencing difficulty, and that imagery is the representation being employed to overcome the difficulty. An illustration of this can be seen in Table 2 below.

Table 2. Category of cognitive procedures and initiation of image production

Category of Cognitive procedure associated with commencement of each sketch episode										
GE = Generation, EX = Exploration, EC = Executive Control											
Student										
Sketching episode	1	2	3	4	5	6	7			
Category of Initial procedure	EC	EC	EC	EX	GE	GE	GE			
Sketch number	1	1	2	2	2	2	2			
Competent architect										
Sketching episode	1	2	3	4	5	6	7	8	9	10	11
Category of Initial procedure	EC	EC	EC	EC	EC	EC	EC	EC	EC	EC	GE
Sketch number	1	2	3	4	5	6	5	6	7	8	9
Expert architect										
Sketching episode	1	2	3	4	5					
Category of Initial procedure	EC	GE	EC	EX	GE					
Sketch number	1	1	2	1	1					

One can conclude from the data in Table 2, that visual mental images are important in facilitating generative, exploratory and executive control procedures, and in turn, in being initiated by those procedures. Furthermore, as expertise

EXAMINING DESIGN THINKING

develops, images are more likely to be initiated by generative than exploratory or executive control procedures.

Secondly, an analysis was made of the dominant procedure occurring with each sketch. This analysis was undertaken to provide information on the progression from problem zone to satisfying zone, and the role of imagery in that process. That is, at what point or points is generation the dominant mode, or exploration, or executive control. An illustration of the representation of the data for this analysis can be seen in Table 3 below.

Table 3. Relationship between cognitive procedures and imagery usage

Most frequently used category of cognitive procedure associated with each sketch episode										
GE = Generation, EX = Exploration, EC = Executive Control										
Student S4										
Sketching episode	1	2	3	4	5	6	7			
Category of procedure	EC	GE	EC	GE	GE	GE/EX	EX			
Sketch number	1	1	2	2	2	2	2			
Competent architect										
Sketching episode	1	2	3	4	5	6	7	8	9	10	11
Most frequently used procedure	GE EX	GE	GE	EC	GE	EX, EC	EC	EX	GE	GE	GE
Sketch number	1	2	3	4	5	6	5	6	7	8	9
Expert architect										
Sketching episode	1	2	3	4	5					
Most frequently used procedure	EX	GE	GE	GE	GE					
Sketch number	1	1	2	1	1					

From the data in Table 3 it is possible to conclude that, firstly, the general pattern of relations between imagery usage and most frequently used cognitive procedure for the student was consistent with that of a problem-solver who had a limited amount of readily applicable procedural knowledge and was encountering some difficulty in solving the problem. In terms of the architects, the expert's activity in terms of most frequently used category of procedures associated with sketch episodes, suggest expert-like features. That is, the first sketch episode for the expert was associated with exploratory categories of procedures. Then, all remaining sketch episodes were associated with generative procedures.

STRENGTHS AND LIMITATIONS OF THE RESEARCH DESIGN

Strengths

The strengths of the research design can be summarised in terms of five features that relate to the data and the research settings. Firstly, the research design ensured that a rich data source was generated and collected. Secondly, the data source consisted of three types of data: verbal protocols; visual protocols; and the video-recording of movement and sound. The collection of three different kinds of data about the same activity made triangulation of coding possible and enhanced internal validity. Thirdly, the method of data collection and analysis was transparent. Fourthly, the activities and settings in which the data were collected represented a high level of authenticity in terms of their resemblance to the activities they were intended to research. Fifthly, the study consisted of two different kinds of subjects (students and architects) in two different kinds of settings, which assists in interpretations of the data.

Internal validity

Internal validity is concerned with whether a study is reporting what it claims to be reporting. There are two main threats to the internal validity of the study. Firstly, there can be problems in deciding before the event, what is to be investigated, because it can lead to important phenomena being overlooked (Jacob, 1990). Secondly, the analysis of cognitive processes in real time is one of the methodologically difficult tasks in psychology. This difficulty can be expected to increase where the various cognitive processes being investigated are likely to be operating in parallel (Olson, Duffy & Mack, 1984).

The degree to which the first threat to internal validity applies depends on the nature of the data collected and the degree of focus of the investigation. That is, if limited data were collected from a small section of a problem-solving performance, it is possible that important phenomena could be overlooked. In the two studies reported in this chapter, the problem-solving performance was examined in totality, using visual and verbal protocols to produce what Simon and Kaplan (1989) would describe as a rich data source. It is therefore argued that while decisions concerning what to investigate were made before the study commenced, the methodology used ensured that important, but not pre-specified phenomena were not overlooked. This concern was also addressed by trialling the design problem with one architect and two students, prior to data collection. This led to refinement of the data collection process to ensure all appropriate data were collected. Further, as the methodology used is a highly descriptive one, the processes of data collection and analysis are particularly transparent and, therefore, all phenomena are visible to the researcher and reader.

In addressing the threat that is encountered by all studies examining internal mental processes, particularly those looking at parallel processing of information, two replies are offered. Firstly, the study relies on 'think-aloud protocols' which provide a rich data source, which, while not constituting cognition as such, do provide important information from which valid conclusions about cognitive processes can be made (Ericsson & Simon, 1993). Secondly, there is the problem of collecting parallel data sources when using the inherently serial data provided by verbal protocols. This problem was addressed by using visual and verbal protocols, thus providing two parallel data sources. Thus, the data collected are both rich and capable of representing parallel cognitive processing. Further, as the task being examined constituted a complex task that all subjects found difficult, the study was likely to collect data concerning higher order cognitive processes.

Another threat to internal validity is caused by the fact that verbal protocols may not be complete, and those verbalisations that are produced may not provide any clues to indicate that data are missing (Duncker, 1945). The threat was addressed in the study by using multiple sources of data. The use of multiple sources of data provided the capacity to examine the transcripts of the verbal protocols and the concurrent visual protocols in any part of the data records. By using both protocol types, gaps in verbalisations can be correlated with the visual protocol. In this way it was possible to 'fill in the gaps' by, for example, indicating that while at a particular stage no verbalisations were produced, the problem-solver was producing sketches, indicating that at that stage, thinking may have been of a predominantly visual nature.

External validity

External validity is concerned with the extent to which the findings of one study conducted in one setting with a particular population can be generalised to other populations or settings (Burns, 1990). It needs to be noted that, as Firestone (1993) argues, the extent to which findings from any one study can be generalised and applied as universal laws with applicability across populations and settings is always problematic. Simon (1975) goes further in cautioning against the use of traditional statistical methods for achieving external validity when arguing for the importance of the study of individual behaviour:

> If we are to understand human problem-solving behaviour, we must get a solid grip on the strategies that underlie that behaviour, and we must avoid blending together in a statistical stew quite diverse problem-solving behaviours whose real significance is lost in the averaging process
>
> (Simon, 1975, 288)

Burns (1990) advances four concerns about external validity. The four concerns are the failure to describe independent variables, the representativeness of the sample, Hawthorne-type effects, and the validity of transferring results from experimental or quasi-experimental settings to other settings. The four concerns are dealt with below within a framework informed by Firestone's (1993) general caution about external validity and Simon's (1975) more specific caution about traditional methods for achieving external validity when examining human problem-solving behaviour. Burns' (1990) concerns about external validity are addressed in the following manner.

Firstly, Burns (1990) suggests one threat to external validity is the failure to describe independent variables. This threat has been addressed through the detailed descriptions and analyses of three subjects in three settings and through the methods used to collect data. By providing a rich data source and detailed descriptions of methods and procedures it is possible for other researchers to replicate the approach and determine the extent to which this study has similarities to their studies.

Secondly, Burns (1990) suggests external validity is threatened if it is not possible to demonstrate that the sample used in the study is representative of the population. The degree to which the sample is representative of designers in general, or architectural designers, or design students, is difficult to establish, without replicating the study with more subjects in other settings. In any case, no claims for generalisation are made from this study. The study is essentially of the complex thinking of a number of subjects. For those who wish to speculate on the potential for generalisation, it could be noted that there is no evidence to suggest that the architects were different in any systematic way from architects in general or even designers, in general; or that the student subjects were systematically different from other high school students involved in learning that involves designing. However, the study seeks only to describe the thinking of these three subjects and to use this as confirmation of the theorising advanced in this study. If the theorising is supported by those three subjects, then the theoretical ideas can be explored further with other subjects in other areas.

Thirdly, Burns (1990) advances concerns about Hawthorne-type effects, where participation in an activity influences the outcomes of the activity. The effect of the participation can be expected to derive from two aspects of the activities. The first is the nature of the activity itself and the second is the presence of the experimenter. The issue of the activity was addressed in the first study by ensuring that the activity was closely related to the classroom activities that the student was engaged in. That is, the class was covering a unit of work on architectural design that involved the solving of a design problem similar to the one used in the study, at the time the data were collected. In the second study, the issue was addressed by developing the task with an architect, accustomed to the kind of task to be used, but not involved in the study.

The presence of the experimenter probably had some effect on the subjects. However, any effect was ameliorated by selecting, for the first study, architects who had experience in domestic architecture and were accustomed to thinking out aloud in front of clients while creating concept sketches of houses of the kind required in the task. In the case of the students, the Hawthorne-type effect was addressed by using only volunteers, and by building into the data collection process an orientation phase. Hawthorne-type threats to external validity are therefore, not seen as a serious threat to this study. Moreover, it was thought that any Hawthorne effect would be unable to press subjects into thinking that was uncharacteristic or beyond their abilities.

Burns' (1990) final concern is the degree to which judgements made in experimental or quasi-experimental settings can be transferred to other settings. This is not seen as a threat to validity in this study, as the architects were engaged in a bona-fide architectural problem in their normal work setting. In the case of the student, it was necessary to remove the subject to a quiet room in which the design problem was attempted and in which it was possible to record the verbalisations of thought and the concurrent visual and motor activity. It is possible that the need to remove the student may have affected the performance on the task and therefore the validity of transferring results from this setting to other settings. However, as previously indicated, only volunteer students were used for the case study, and the design problem being attempted was sufficiently complex to require considerable attention from the subject. Thus, while the results of the student study may have been affected by the research setting, the effect is regarded as not having posed any serious threat to validity.

Nevertheless, no claims are made that the results of this study are generalisable. Rather the conclusions constitute an illumination of the kinds of thinking that was engaged in by these subjects. It is acknowledged that considerable further work would be required to confirm the results more generally.

Reliability

The reliability of any research is the degree to which the results of the research would be replicated by another researcher examining the same phenomena. Reliability can be divided into internal and external reliability. Internal reliability, is the degree to which other researchers, using the theory and constructs of the study, would achieve the same results. External reliability is the extent to which an independent researcher, examining the same phenomena, would come up with the same constructs and results.

Reliability is an important factor that affects the credibility of any research. Nevertheless, with qualitative research, the ability to reproduce any study exactly is low (Le Compte & Goertz, 1982). In this study, none of the problem-solving episodes could have been reproduced. To reduce concern about reliability two approaches were used. External reliability was addressed by providing detailed descriptions of settings, subjects, procedures and methodologies in order to give other researchers comprehensive information to use in seeking to undertake similar studies. Internal reliability is addressed below in terms of coding reliability by establishing measures of inter-coder reliability.

Internal reliability

Coding reliability. A measure of coding reliability was established by coding a sample of the protocols using an independent rater. The independent rating was conducted by providing the independent rater with a copy of the coding conventions outlined above. The rater was instructed via discussion of the codes and by coding a sample of protocol with the researcher. After discussion the rater independently coded a section of one protocol. This illuminated the need for greater clarity in terms of the coding descriptors. After clarifying remaining areas of the coding scheme, the rater coded all selected protocol samples.

After coding all sample protocols, protocols where agreement was not achieved, were examined. To resolve disagreement, the verbal protocols were examined in terms of the information contained in the visual protocols and the video-recording. The result of this examination was an increase in the level of agreement between raters. A high degree of congruence between the researcher's coding and that of the independent rater was achieved over three hundred and eighty-eight protocol segments. In Table 3, the level of congruence across all categories of subjects selected for establishing coding reliability is provided.

Table 3 *Frequency of congruence between raters on sample protocols.*

Number of items coded	Percentage agreement between researcher and independent rater
127	91%
85	96%
80	89%
Total = 292	Total = 92%

Protocol segments where agreement between raters was not achieved after discussion and reference to visual protocols and video-recording data were examined. The conclusion was reached that lack of coding agreement was the result of ambiguities in the semantic content of the protocols which provided the basis for different interpretations of meaning and thus, different assignment of codes. However, given the nature of the tasks, the level of congruence achieved was regarded as acceptable.

Limitations

For practical reasons it was necessary to restrict the studies to architectural designers, engaged in domestic architecture, and a design student. This first limitation, in breadth, is offset by the degree of depth achieved, in terms of the analysis of individual problem-solving performance. A number of other limitations of the study can also be identified.

Firstly, while the use of verbal protocols in the study provided a valuable method for collecting data about knowledge that subjects were attending to, verbal protocol analysis is unable to elicit problem-solving information that is in a tacit or automated form. The limitations of verbal protocols to illustrate tacit knowledge has been addressed in two ways: firstly, by using a task that was intended to be difficult for all subjects to solve, it was posited that solutions would not be achieved through access to significant tacit knowledge; and secondly, Finke (1989) has demonstrated that many forms of tacit knowledge can be accessed only from long-term memory with imagery processes. Therefore, by using visual as well as verbal protocols, it is argued that there is a high probability that any tacit knowledge that was used would have manifested itself in sketches and be captured in visual protocols.

Secondly, while every effort was made to ensure that the problem-solving task was seen as authentic by each subject and the settings were as natural as possible, it is acknowledged that the tasks were not in fact totally authentic, as far as the architects were concerned, given that the solutions would not be built nor the designs paid for. In terms of the students, the need to remove the subject from the design classroom to an adjacent room in order to collect verbal protocols, probably also affected the degree to which the task was seen as authentic. Nevertheless, the richness of the data suggests the tasks were performed as if they were authentic activities.

CONCLUSIONS

This chapter provides a description of an examination of design thinking using verbal and visual data. Initially, the analysis of verbal data is undertaken using standard protocol analysis methods. This provided findings on how people deployed cognitive procedures when designing. Then, an analysis of the use of visual data was described. Finally, a description of the analysis of the combined data was provided. This produced findings additional to those produced by the separate analyses of visual and verbal data.

It is argued that the strength of the research design described, resides in its use of tasks and settings that have a high degree of authenticity, the richness of the data provided by the complementary verbal and visual protocols and the linking of theoretical frameworks to the research design and procedures which, while having direction, allow the data to influence findings.

REFERENCES

Anderson, J. R., Qin, Y., Jung, K. J., & Carter, C. S. (2006). Information-processing modules and their relative modality specificity. *Cognitive Psychology*.

Anderson, J. R. (1993). Problem solving and learning. *American Psychologist, 48*, (1), 35-44.

Burns, R. (1990). *Introduction to research methods in education*. Sydney: Longman Cheshire.

Chan, C. S. (1990). Cognitive processes in architectural design problem solving. *Design Studies, 11*, (2), 60-80.

Dreyfus, H. L., & Dreyfus, S. E. (1986). *Mind over machine: The power of human intuition and expertise in the era of the computer*. New York: The Free Press.

Duncker, K. (1945). On problem solving. *Psychological Monographs, 5*, (All No. 270).

Eckersley, M. (1988). The form of design processes: A protocol analysis study. *Design Studies, 9*(2), 86-94.

Ericsson, K. A., & Simon, H. A. (1993). *Protocol analysis: Verbal reports as data*. Cambridge MA: MIT Press.

Finke, R. A. (1989). *Principles of mental imagery*. Cambridge MA: MIT Press.

Finke, R. A., Ward, T. B., & Smith, S. M. (1992). *Creative cognition*. Cambridge, MA: MIT Press.

Firestone, W. A. (1993). Alternative arguments for generalising from data as applied to qualitative research. *Educational Researcher, 22*, (4), 16-23.

Gay, L. R. (1993). *Educational research: Competencies for analysis and application*, (3rd Edition). New York: Macmillan.

Gross, M., & Fleisher, A. (1984). Design as the exploration of constraints. *Design Studies, 5*, (3), 137-138.

Jacob, E. (1990). Alternative approaches for studying naturally occurring human behaviour and thought in special education research. *The Journal of Special Education, 24*, (2), 195-211.

Kosslyn, S. M. (1994). *Image and brain: The resolution of the imagery debate*. Cambridge, MA: Bradford Books.

Larkin, J. H. & Simon, H. A. (1987). Why a diagram is (sometimes) worth ten thousand words. *Cognitive Science, 11*, 65-99.

Le Compte, M. D., & Goertz, J. P. (1982). Problems of reliability and validity in ethnographic research. *Review of Educational Research, 52*, (1), 31-60.

Middleton, H. E. (1998). The role of visual mental imagery in solving complex problems in design. Unpublished PhD dissertation. Griffith University.

Olson, G. M., Duffy, S. A., & Mack, R. L. (1984). Thinking-out-aloud as a method for studying real-time comprehension processes. In D. E. Kieras & M. A. Just (Eds), *New methods in reading comprehension research*. Hillsdale, NJ: Erlbaum. 253-286.

Onions, C. T. (1974). *The shorter oxford english dictionary on historical principles*. Oxford: The Clarendon Press.

Pylyshyn, Z. W. (1981). The imagery debate: Analogue media versus tacit knowledge. *Psychological Review, 88* (1), 16-45.

Schon, D. A. (1990). The design process. In V. A. Howard (Ed), *Varieties of thinking: Essays from Harvad's philosophy of education research centre*. New York: Routledge.

Shepard, R. M. (1978). Externalization of mental images and the act of creation. In B. S. Randhawa and W. E. Coffman (Eds.) *Visual learning, thinking and communication*. New York: Academic Press.

Simon, H. A. (1975). The functional equivalence of problem solving skills. *Cognitive Psychology, 7*, 268-288.

Simon, H. A. (1973). The structure of ill-structured problems. *Artificial Intelligence, 4*, 181-201.

Simon, H. A., & Kaplan, C. A. (1989). Foundations of cognitive science. In M. I. Posner (Ed.), *Foundations of Cognitive Science*. Cambridge, MA: MIT Press.

Smith, S. M., Ward, T. B., & Finke, R. A. (1995). Cognitive processes in creative contexts. In S. M. Smith, T. B. Ward & R. A. Finke (Eds.) *The creative cognition approach*. Cambridge, MA: Bradford Books (2-5).

Weber, R. J., Moder, C. L., & Solie, J. B. (1990). Invention heuristics and mental processes underlying the development of a patent for the application of herbicides. *New Ideas in Psychology, 3*, 321-336.

Yin, R. K. (1994). *Case study research: Design and methods*. Thousand Oaks: Sage Publishing.

Howard Middleton
Griffith Institute for Educational Research
Griffith University
Australia

INDEX

—3—
3D-CAD, 69, 70, 71, 72, 74, 75, 76, 78, 79, 85

—A—
Ability, 63
Accuracy, 57
Action, 25, 160, 161
Action research, 3, 9, 13
Action research approach, 9
Actions, 160, 162
Activity, 157, 167, 177, 188
Advanced, 67
Aesthetic, 57
Analyses, 3, 14, 15, 22, 23, 62, 117, 120, 124, 203, 206, 210
Analysis, 2, 3, 4, 12, 14, 15, 16, 20, 22, 23, 24, 26, 28, 36, 44, 54, 55, 56, 61, 65, 69, 70, 71, 72, 74, 75, 76, 78, 79, 80, 82, 83, 84, 85, 86, 87, 95, 97, 98, 110, 111, 114, 115, 116, 117, 118, 119, 120, 121, 122, 123, 124, 125, 126, 127, 128, 129, 130, 131, 132, 137, 145, 151, 156, 160, 171, 176, 177, 181, 182, 184, 185 186, 191, 192, 194, 197, 198, 200, 201, 203, 204, 205, 209, 210
Analytic generalisation, 11
Approach, 66, 67, 97, 156, 189
Approaches, 25, 134, 135, 152, 153, 187
APU D&T project, 134, 135
Art, 5, 46, 47, 57, 62, 132, 152, 187
Artefacts, 160
Assessing, 25, 46, 62, 67, 113, 150, 152
Assessing design innovation, 67
Assessment, 25, 26, 67, 98, 134, 153, 189
Audio, 18, 19, 20, 22, 51, 106, 157, 160, 166, 168, 177, 178, 181, 182
Audio and video recording, 25
Audio recording, 4, 13, 17, 18, 19, 20, 22, 74, 76, 77, 78, 85, 95, 129, 138, 140, 156, 157, 158, 177, 179, 180, 181, 182, 185, 194

Australia, 5, 44, 67, 97, 117, 127, 128, 132, 133, 155, 162, 169, 173, 176, 188, 189, 211
Authentic, 186

—B—
Beautifulness, 56
Bias, 7, 11, 13, 14, 15, 16, 61, 65

—C—
CAD, 83
CAD, 69, 70, 71, 74, 76, 78, 79, 80, 82, 83, 84, 85, 86
Camtasia, 77, 78, 79
Canada, 28, 38, 40
Cartesian system, 55
Case studies, 2, 25
Case study, 2, 3, 6, 7, 8, 9, 10, 11, 12, 13, 15, 22, 24, 25, 26, 28, 29, 30, 36, 38, 43, 44, 67, 134, 175, 192, 208, 211
Case Study Methodology, 7
Categories, 195
Causal studies, 16
Checking, 83, 84
Children's concepts, 91
Children's thinking, 91, 95
Choice corollary, 61
Circuit, 23
Circuits, 17, 22, 30, 69
Coding, 82, 137, 153, 195, 199, 208
Coding schemes, 137
Cognitive, 9, 26, 48, 66, 67, 80, 82, 84, 86, 87, 174, 176, 177, 187, 189, 203, 210, 211
Cognitive Acceleration Programme, 9
Communication, 159
Comparative, 109, 114, 115, 119, 125, 131, 132, 133
Comparative judgement, 114
Comparative Pairs test, 109
Competent, 197, 203, 204
Complex, 67, 169
Component, 54, 55, 56
Components, 55
Concept, 3, 7, 11, 23, 29, 33, 49, 64, 88, 91, 92, 97, 99, 102, 103, 104, 111,

213

113, 117, 118, 120, 121, 125, 127, 129, 130, 131, 168, 207
Concept formation, 3, 88, 90, 91, 94
Concepts, 3, 6, 8, 10, 12, 16, 23, 25, 38, 41, 42, 60, 88, 91, 92, 93, 94, 95, 103, 120, 121, 128, 163, 165, 166
Concepts, 66
Conceptual, 6, 12, 23, 64, 91, 95, 97, 117, 118, 130, 152, 166, 169, 183
Conceptual connections, 95
Conceptual knowledge, 6, 23
Conceptual tools, 91
Conductivity, 7
Conductors, 22, 30, 31
Construct, 12, 35, 47, 49, 58, 60, 61, 66, 165
Construct validity, 11
Constructing, 22
Construction, 82, 83, 84
Constructivist, 66, 67, 88, 89, 90
Constructs, 49, 59
Content, 32, 33, 44, 124, 187, 188
Context, 4, 6, 7, 8, 9, 10, 20, 28, 29, 34, 36, 38, 42, 43, 50, 65, 69, 89, 90, 92, 93, 95, 96, 103, 107, 115, 117, 118, 119, 120, 123, 130, 131, 139, 149, 156, 158, 168, 169, 170, 196
Craft, 46, 67
Craftmanship, 56
Creating, 141, 160, 163, 165, 188
Creative, 46, 52, 56, 57, 67, 86, 87, 104, 210
Creativity, 62, 67
Critical, 116
Cultural historical theory, 89
Cultural-historical, 3, 4, 88, 89, 92, 93, 94, 97
Cultural-historical methodology, 95
Cultural-historical perspective, 88, 89, 93
Cultural-historical psychology, 90
Cultural-historical research, 94, 97
Cultural-historical tools, 92
Culture, 188
Current state, 143
Curriculum documents, 126

—D—

Data analysis, 2, 11, 77, 137, 170, 178, 184, 185

Data collection, 7, 9, 11, 12, 14, 15, 18, 20, 58, 71, 74, 76, 77, 79, 128, 156, 175, 180, 182, 186, 204, 205, 207
Data collection, 76, 79
Data collection techniques, 170
Data segmentation, 80
Data selection, 22
DEPTH, 28, 38, 42, 43, 44
Design, 25, 29, 33, 34, 36, 42, 44, 46, 62, 64, 66, 67, 68, 70, 86, 101, 102, 103, 113, 114, 132, 134, 136, 150, 152, 153, 187, 188, 189, 192, 193, 210, 211
Design and Technology, 25, 29, 34, 42, 44, 46, 67, 96, 113, 114, 134, 153, 187, 188, 189, 193
Design experiment approach, 9
Design Experiments, 136
Design process, 6, 19, 46, 47, 51, 64, 85, 137, 171, 211
Design project, 101
Designers, 25
Designing, 143, 188, 191
Development, 32, 38, 44, 66, 97, 169, 188, 189
Developmental, 66
Dialogue, 26
Differentiation-polarization theory', 8
Direction, 23, 146
Discussions, 102, 157
Double move approach, 88

—E—

E-learning, 102
Electrical, 22, 31, 32
Electrics, 9
Electronic, 9, 22, 50, 69, 102
Electronic activities, 9
Electronics, 6, 8, 11, 17, 20, 21, 23, 26, 30, 33
Elements, 49, 51, 54, 58, 62
Elicitation, 71, 86
Eliciting, 59, 66, 86
Emergent pole, 52
Empirical knowledge, 92
Empirical psychology, 90
End product, 36, 88
Energy and the Environment, 150, 152
Engaged, 57
Engagement, 57
Epistemological, 25, 26

E-scape, 113
Ethics, 7, 11, 13, 17, 26, 92, 142
Ethnographic, 188
Ethnographies, 10
Ethnography, 10
Evaluation, 26, 83, 84, 133, 195
Evidence, 66
Executive control, 195
Experience, 61
Experiential learning, 171
Experimental, 67, 86, 187
Expert, 48, 59, 66, 203, 204
Expertise, 66, 67, 69, 70, 189
Experts, 47, 188
Explanatory, 9, 16, 24
Exploration, 195, 196, 203, 204
Exploring attributes, 195
Exploring constraints, 195
External validity, 36, 206

—F—

Familiarity, 143
Fantastic, 57
Feminism, 116
Field experiment, 9
Finland, 28, 38, 44
Focus, 54, 188
Focus display, 54
Framework, 25, 26, 36, 38, 97, 171, 188
Function, 165, 202
Functional, 165
Functionality, 53, 56, 57
Fundamental Postulate, 61

—G—

GCSE, 102, 103, 110, 112, 114, 135
Gender, 11, 26, 107, 143
Generalisability, 11
Generalisation, 7, 10, 11, 207
Generalise, 11
Generating, 86
Generation, 195, 203, 204
Genlock, 157, 160
Geometry Identification, 82, 84
Goal setting, 195
Goal switching, 195
Graphics, 37

—H—

Hawthorne, 207
Hawthorne-type effects, 207
High grade, 52, 54, 57

Higher-order, 170, 171, 173, 174, 175, 176, 178, 183, 184, 187
Higher-order thinking, 170, 173, 174, 175, 176, 178, 183, 184, 187
Higher-order thinking skills, 171
Highly innovative, 57
Historical dimensions, 93

—I—

Ideas, 86, 211
Ideas technology, 88, 91, 92
Identifying, 137
Illustration, 8, 31, 58, 191, 202, 203, 204
Image type, 202
Imagery, 86
Implicit, 52, 66, 67, 118
Information, 66, 113, 114, 210
Innovation, 113, 150, 152, 189
Insight, 66
Instruction, 161, 169, 177, 187, 189
Instructional strategies, 171, 172
Instructional strategy, 171, 172
Instructional tactics, 172, 174
Intentions, 188
Interaction, 66, 67
Internal, 15, 16, 205, 208
Internal validity, 11, 205
Interview, 13, 20, 21, 39, 46, 48, 49, 50, 51, 58, 60, 62, 65, 72, 75, 94, 125, 128, 129, 135, 168, 170, 174, 175, 176, 178, 179, 180, 181, 182, 183, 184, 185, 186
Interview schedule, 20
Interviewing, 14, 18, 20, 129, 170, 178, 187
Interviews, 3, 9, 12, 18, 20, 21, 22, 24, 29, 32, 62, 64, 71, 75, 76, 95, 119, 122, 124, 125, 128, 129, 130, 131, 143, 157, 160, 166, 170, 175, 177, 178, 179, 180, 181, 182, 183, 184, 185, 186, 187
Intra-judge reliability, 13
Introspection, 86
Intuition, 66, 67
Inventiveness, 56, 57, 62
Investigating, 67

—K—

Key Stage, 25, 145, 146
Key Stage performance, 146

INDEX

Knowledge, 1, 2, 3, 4, 5, 6, 7, 8, 9, 12, 13, 17, 18, 22, 23, 24, 25, 26, 28, 29, 30, 31, 32, 33, 34, 35, 36, 37, 38, 39, 40, 41, 42, 43, 44, 46, 47, 48, 49, 58, 61, 64, 65, 66, 67, 69, 70, 71, 72, 74, 75, 76, 77, 78, 79, 82, 84, 85, 86, 88, 89, 94, 95, 97, 106, 117, 118, 120, 121, 122, 123, 125, 126, 127, 128, 129, 130, 131, 132, 154, 155, 160, 162, 163, 164, 165, 166, 167, 168, 169, 170, 171, 175, 176, 184, 185, 187, 188, 191, 192, 195, 204, 209, 211
Knowledge generation, 91
Kunst, 120

—L—

La transposition didactique, 34
Learning environments, 170, 173, 176, 177, 183, 186, 187
LEGO Education, 150
Literacy, 25, 26, 187
Looking, 104, 145, 148
Loose endings, 57
Low grade, 52, 55, 57

—M—

Making, 46, 107, 189
Manageability, 112
Manifestations, 148
Mathematics, 6, 10, 12, 33, 98
Meaning, 158
Meanings, 158
Mediated action, 90
mediational means, 90
Meeting the standards, 56
Mental subtraction, 80
Metacognitive, 84
Metaphors, 13
Methodological, 25, 89, 156
Methodological individualism, 89
Methodologies, 26, 93, 95
Methodology, 68, 78, 174, 187, 188, 189
Microsoft Excel spreadsheet, 144
Modelling, 86
Monitoring, 83, 84, 195
Motoring, 145

—N—

Narrative accounts, 25
National curriculum, 50

National Curriculum, 102, 113, 134, 144, 145, 189
Nature of knowledge, 4, 6
Nauka, 120
Needs support, 52, 54, 56
Neo-liberalism/Neo-Managerialism, 116
New Zealand, 28, 38, 44, 170, 188
No Endurance, 52, 56
Novice, 52, 59

—O—

Observation, 136
Observational, 134
Observations, 4, 156, 157
Observer, 61, 168
Observing, 140
Operation, 161
Operations, 158
Order, 127, 189
Outcomes, 66, 188

—P—

Participant observation, 10, 14, 16, 25, 136, 141
Participant observation, 136
Past experience, 143
Past experiences, 143
PATT, 1, 28, 189
Pedagogic, 44, 189
Pedagogical, 33, 37, 41, 44
Pedagogical content knowing, 44
Pedagogy, 6, 33, 34, 44, 67, 97, 131, 146
Perception, 86
Performance, 25, 66, 98, 134, 153, 188, 189
Persistent, 52, 54, 56, 57
Personal subject construct, 37, 42
Personal/Construct, 60, 61
Phenomena, 2, 4, 7, 9, 10, 28, 48, 124, 131, 167, 205, 208
Physiological, 143
Planning, 83, 84
Post-colonialism, 116
Post-modernism, 116
Post-structuralism, 116
Predicting, 83, 84
Preparing, 169
Primary, 26
Printed circuit board, 9
Problem solving, 6, 8, 9, 11, 13, 17, 22, 23, 25, 26, 41, 46, 47, 67, 70, 72, 73,

216

INDEX

75, 86, 87, 155, 157, 169, 177, 210, 211
Problems, 25, 68, 87, 97, 133, 169, 211
Problem-solving, 188
Procedural, 6, 12, 23, 25, 26, 47, 48, 60, 75, 78, 84, 85, 99, 100, 104, 151, 191, 204
Procedural, 188
Process, 62, 66, 67, 76, 78, 84, 189
Processes, 3, 6, 11, 23, 25, 36, 43, 46, 48, 60, 61, 69, 71, 72, 73, 74, 76, 78, 79, 80, 82, 83, 84, 85, 86, 88, 89, 95, 98, 100, 101, 102, 103, 120, 121, 131, 136, 137, 151, 154, 158, 159, 168, 170, 173, 175, 176, 184, 185, 188, 193, 198, 205, 209, 210, 211
Progress, 133, 169
Properties, 202
Protocol segments, 209
Protocols, 202

—Q—

Qualitative, 2, 6, 8, 10, 18, 24, 48, 57, 60, 62, 65, 95, 97, 115, 119, 122, 124, 128, 130, 131, 132, 137, 141, 152, 174, 175, 181, 186, 187, 189, 208, 210, 211
Qualitative data, 10, 137, 141, 187
Qualitative knowledge, 6
Quality, 11, 19, 20, 22, 24, 37, 41, 49, 63, 103, 107, 108, 109, 111, 115, 124, 137, 142, 149, 151, 185, 186
Quantifying, 66
Quantitative, 10, 13, 18, 23, 33, 60, 62, 115, 131, 136, 137, 141, 174, 175, 186, 187, 192, 211
Quantitative analyses, 13
Quantitative data, 18, 60, 62, 136, 141
Quasi-experimental, 9, 206, 207

—R—

Reading, 114
Recording, 189
Reduction, 22
Reflective, 60, 67, 187
Relevance, 21
Relevant-Irrelevant, 59
Reliability, 11, 13, 22, 57, 60, 65, 66, 110, 111, 112, 130, 136, 142, 143, 149, 178, 208, 209, 211
Reliable, 57

Reliable data, 21
Repertory Grid, 46, 47, 49, 50, 51, 58, 60, 61, 62, 64, 65, 66, 67
Repertory Grids, 66
RepGrid IV, 50
Representation, 87, 200
Research design, 11, 90, 186, 204, 210
Research methods, 66, 92, 93, 94
Researcher, 25, 152, 188, 210, 211
Researchers, 95, 168
Researching, 5, 69, 88, 95, 113, 153
Resistor, 9, 24
Retrieval, 195
Retrospective reports, 74
Reviewing, 153
Rotation sweep, 81
Russia, 117, 120, 121, 127, 128, 129, 132

—S—

Sample, 11, 17, 18, 22, 98, 111, 112, 134, 141, 143, 144, 150, 175, 206, 207, 208, 209
Sample lessons, 17
Sampling, 11, 17, 18, 19, 51, 65, 175
Scatterplot, 198, 199
Science, 6, 7, 8, 9, 12, 16, 22, 25, 26, 30, 33, 34, 47, 48, 49, 50, 61, 64, 66, 68, 82, 86, 90, 92, 97, 98, 117, 120, 132, 149, 150, 155, 169, 187, 188, 211
Science knowledge, 6, 8, 9, 16, 22
Search, 189, 190
Secondary, 103
Selecting, 188
Self confident, 56
Sense, 132, 153
Situated, 6, 7, 48, 90
Sketch, 202, 203, 204
Skill, 67, 169
Skills, 67, 101, 102, 113, 114, 152, 155, 169
Sloppy, 57
SMEs, 77
Social or moral issues, 6
Sociological, 25
Spatial, 83, 84
Standards, 114, 127, 132, 187
Statistical generalisation, 10, 11
Statistical methods, 54
Stimulated recall, 4
Strategies, 161, 188

217

INDEX

Strategy formulation, 195
Structure, 87, 188
Structured, 75
Structured interviews, 20, 75, 157
Student questionnaires, 9
Studies, 6, 7, 8, 10, 11, 13, 22, 24, 25, 30, 40, 43, 46, 47, 51, 58, 61, 62, 63, 64, 90, 108, 112, 115, 116, 120, 126, 130, 168, 176, 189, 191, 192, 194, 202, 205, 206, 208, 209
Subjects, 79, 194
Support, 146
Symbiotic relationship, 91
Synthesis, 86, 195
Systematic observation, 136
Systems, 17, 22, 48, 49, 60, 66, 71, 78, 85, 86, 93, 97, 100, 104, 108, 116, 117, 118, 119, 132, 137, 144, 154, 156, 169, 170, 171, 172, 175
Systems, 48, 66, 67

—T—

Tacit, 46, 47, 48, 64, 67
Task, 79, 87
Teacher's role, 6
Technik, 120, 121, 131, 133
Technique, 46, 49, 58, 60, 61, 62, 64, 65, 66, 67, 177
Techniques, 58, 134
Technological, 25, 26, 66, 97, 132, 134, 135, 152, 187
Technological literacy, 155, 171
Technology, 5, 25, 26, 28, 29, 33, 44, 45, 46, 57, 66, 67, 68, 69, 91, 92, 97, 98, 101, 103, 113, 117, 127, 128, 131, 132, 133, 134, 150, 152, 153, 171, 173, 183, 187, 188, 189, 190
Technology activity, 6, 10
Technology education, 1, 2, 3, 4, 11, 18, 22, 24, 26, 39, 65, 88, 89, 90, 91, 92, 96, 97, 115, 116, 117, 118, 119, 121, 122, 123, 126, 127, 128, 129, 130, 131, 132, 136, 150, 151, 153, 154, 170, 171, 172, 173, 174, 175, 176, 177, 182, 183m 184, 186, 187, 188
Technology knowledge, 17, 155
Technology Postgraduate Certificate in Education, 29
Tekhnika, 121
Tekhnologiya/Technologie, 121
Teknik, 46, 50

Tentiles, 197
TERU, 25, 98, 99, 102, 103, 106, 113, 134, 150
The *use of knowledge*, 6, 17
Theoretical generalisations, 24
Theoretical knowledge, 91
Theoretical model, 10, 129
Theoretical orientation, 96
Theories, 8, 11, 25, 46, 47, 48, 61, 116, 119, 122, 126, 128
Theory, 8, 9, 11, 12, 17, 25, 26, 33, 44, 47, 49, 60, 86, 88, 89, 90, 96, 97, 109, 113, 117, 118, 119, 122, 128, 130, 131, 133, 174, 182, 208, 211
Theory building, 8
Think-aloud, 72, 76
Thinking, 28, 44, 67, 97, 188, 189, 211
Thought, 159, 169
Topics, 132
Traditional, 52, 56, 119
Transcribing, 22
Transcript, 13, 14, 15, 182
Transcription, 13, 15, 20, 22, 75, 178, 181, 182
Transcripts, 22, 206
Transformation, 195
Triangulation, 122, 188
Triangulation of data, 10, 128, 166

—U—

UK, 26, 29, 38, 42, 43, 44, 45, 67, 97, 98, 102, 114, 117, 120, 127, 128, 134, 150, 152
Unengaged, 57
Unit of analysis, 10, 89, 119
United Kingdom, 28, 36
Units of articulation, 73
Unobtrusive methods, 123
Unstructured, 20
Unstructured interview, 20
USA, 86, 97, 117, 127, 128, 189
User, 64, 66
Using technology, 169

—V—

V=IR, 30
Validity, 3, 11, 12, 13, 15, 16, 22, 36, 42, 60, 63, 65, 71, 73, 74, 76, 77, 110, 111, 120, 128, 130, 136, 142, 149, 154, 168, 175, 176, 186, 196, 204, 205, 206, 207, 211

218

Verbal, 72, 80, 210
Video, 4, 13, 14, 15, 17, 18, 19, 20, 21, 22, 24, 58, 76, 77, 78, 79, 80, 82, 85, 95, 151, 157, 160, 166, 168, 174, 175, 176, 177, 178, 179, 180, 181, 182, 183, 184, 185, 186, 187, 193, 194, 204, 208, 209
Video stimulated recall, 170
Video-recording, 13, 14, 15, 19, 194, 204, 208, 209

Visual and audio record, 25
Visual mental images, 199

—W—

WEBGRID III, 50
Wicked, 152
Wissenschaft, 120
Work technologically, 171
Writing, 114